PENGUIN BOOKS

WORKING FIRE

Zac Unger is a firefighter and paramedic in Oakland, ___. ___ about his life and work for the online magazine *Slate*. A graduate of Deep Springs College, Brown University, and UC Berkeley, he lives in Berkeley with his wife and their two children. He has a Web site: www.zacunger.com.

Praise for *Working Fire*

"Unger's memoir gives an unflinching account of the competitive antics, depressing casualties, and often hilarious verbal jousting inside an Oakland ladder company, while also charting his own metamorphosis from naïve Ivy League 'cake-eater' to full-fledged fireman. The book is filled with revelatory nuggets . . . that refreshingly humanize these courageous warriors."
—Raymond Fiore, *Entertainment Weekly*

"The book works on two levels: as an inside view of firefighting that vividly recreates the excitement and fear intrinsic to it, and as an account of how a son of the flower-power class turned into the real thing, a passionately dedicated firefighter. It doesn't hurt that Unger is a lucid writer whose prose almost always is set at just the right pitch. . . . *Working Fire* leaves no doubt that he can do two jobs at once, and do both of them uncommonly well."
—Jonathan Yardley, *The Washington Post*

"[*Working Fire*] sucks you in. And like the best adventure tales—from Jon Krakauer's to Sebastian Junger's—it steps out of the way and lets you go through the adventure yourself." —Daniel Torday, *Esquire*

"*Working Fire* . . . is sometimes moving (his description of the funeral for a fallen firefighter will likely bring tears to your eyes), sometimes funny (he's called to provide medical assistance to a drug dealer who has hidden his wares in a delicate place), but always intelligent and perceptive. . . . *Working Fire* is the best book about firefighting since Dennis Smith's *Report from Engine Co. 82*."
—Curt Schleier, *The Kansas City Star*

"'I can't say for sure if firefighting attracts a certain type of person or if it creates them,' writes Zac Unger, who lays out a solid case for both theories in his *Working Fire: The Making of a Fireman* . . . His account of the work and its gritty allure is masterful." —Jonathan Miles, *Men's Journal*

"All the critical stuff Hollywood never had time to tell you in special effects epics like *Backdraft*. . . . A narrative consistently uplifted by candor and sensitivity . . . Full of rare insights on one of the toughest jobs anyone has to do."
—*Kirkus Reviews* (starred review)

PENGUIN BOOKS

Published by the Penguin Group
Penguin Group (USA) Inc., 375 Hudson Street, New York, New York 10014, U.S.A.
Penguin Group (Canada), 10 Alcorn Avenue, Toronto, Ontario,
Canada M4V 3B2 (a division of Pearson Penguin Canada Inc.)
Penguin Books Ltd, 80 Strand, London WC2R 0RL, England
Penguin Ireland, 25 St Stephen's Green, Dublin 2, Ireland (a division of Penguin Books Ltd)
Penguin Group (Australia), 250 Camberwell Road, Camberwell,
Victoria 3124, Australia (a division of Pearson Australia Group Pty Ltd)
Penguin Books India Pvt Ltd, 11 Community Centre, Panchsheel Park,
New Delhi - 110 017, India
Penguin Group (NZ), cnr Airborne and Rosedale Roads, Albany, Auckland 1310,
New Zealand (a division of Pearson New Zealand Ltd)
Penguin Books (South Africa) (Pty) Ltd, 24 Sturdee Avenue, Rosebank,
Johannesburg 2196, South Africa

Penguin Books Ltd, Registered Offices:
80 Strand, London WC2R 0RL, England

First published in the United States of America by The Penguin Press,
a member of Penguin Group (USA) Inc. 2004
Published in Penguin Books 2005

1   3   5   7   9   10   8   6   4   2

Names have been changed and dialogue is to the best of the author's recollection.

THE LIBRARY OF CONGRESS HAS CATALOGED THE HARDCOVER EDITION AS FOLLOWS:
Working fire : the making of an accidental fireman / Zac Unger.
p.   cm.
ISBN 1-59420-001-7 (hc.)
ISBN 0 14 30.3495 2 (pbk.)
1. Unger, Zac.   2. Firefighters—United States—Biography.   I. Title.
TH9118.U54A3   2004
363.37'092—dc21
[B]   2003050676

Printed in the United States of America
Designed by Stephanie Huntwork

Working Fire

· THE MAKING OF A FIREMAN ·

Zac Unger

PENGUIN BOOKS

For my parents

CONTENTS

Working Fire

· 1 ·

Confirming Stills

In my first months it was all pole. I couldn't imagine taking the stairs like most of the old-timers. For me the pole stood for everything I admired about the fire department—speed, daring, an ageless tradition offered to only a select few. After dark I would lie in my bunk, fully dressed and vibrating with energy, waiting wide-eyed for a call to come in. During my first months in the firehouse, I refused to let myself sleep, fearful that I'd somehow miss the cascading bells and harsh fluorescent lights that erupt when an alarm comes in. When the bells went off, I'd go flying out of my bunk and hit the pole running, spin down to the ground floor, sprint to the rig, and wait. Within thirty seconds—it seemed like an eternity—the others would arrive at the rig, snapping their suspenders into place, kicking sleepy kinks out of their knees. The officer would key the mike to tell dispatch that we were leaving, then flick a match to his cigarette and settle into his seat. We'd race through darkened streets, the officer's tobacco smoke curling back to where I sat fumbling with my heavy coat and gloves. Usually we'd leave the sirens turned off, and I'd watch our progress in the blinking lights playing on the shuttered shop windows and the huddled bodies of men sleeping in doorways. There was no good reason to wake them up.

"Confirming stills. Confirming stills." The voice of a female dispatcher crackles over the radio, telling us that it's likely to be a real fire this time, not just one of our many false alarms. If more than one person calls it in, the flames are apt to be big and showy. They call it a "still" because when the information comes over the firehouse loudspeaker, conversations die in midair, the clink of coffee cups goes silent, and every fireman holds his breath, hoping that the address is in his district and he'll get to go fight a fire. That's the only time a fireman is ever still. A "confirming still" means that the fire is big enough to have attracted attention from several people, all of whom called in to 911 simultaneously.

With the confirmation the officer stubs out the cigarette on the side of his boot and rolls down the window, sniffing for smoke on the quiet summer air. He flips on the siren now, and I hear it echoed by other distant fire engines triangulating their way toward us. "I'm not smelling it yet. You smell anything?" he asks.

"Nope," says Jack Alvarez, the firefighter sitting beside me. "Probably ain't shit." He'd been sleeping, bent forward in his seat, his head resting in the cup of his two hands. I didn't see him put the plug of tobacco into his lower lip. Maybe he goes to bed with it already in place. He rouses himself and slips fluidly into his turnout coat, buckling it over his thick cotton sweatshirt, the one with the logo of a grinning clown holding a fireman's ax. He crosses himself, then kisses his wedding ring, and I make a mental note to develop a ritual for myself, some sort of secular facsimile of prayer that I can use before I go running into a burning building.

"Aw, hell. We'll see it when we get there." The officer rolls up his window and shakes a fresh cigarette out of the package. Alvarez clips the last buckle on his air bottle and then settles his head back down into his hands and closes his eyes.

We turn the last corner, and we're right on top of it. The building on fire is a classic "taxpayer"-type structure, meaning it has apartments on the upper floors and shops at street level. Fire runs the length of the ground floor, shooting out of plate-glass windows shattered by the heat. Alvarez claps his gloved hands and howls like a kid on a roller coaster. Engine Fifteen is already on the scene, and they've hooked a hydrant and laid big-line supply hose clear across the middle of the intersection. What few cars are out at midnight on a Tuesday are driving right over the hose as if they can't see our flashing lights. It's not the first fire I've been to in my few weeks on the job, but I can tell that it's going to be the best one yet.

Truck Four pulls up at the same time we do, slaloming into position nose to nose with our engine. The guys are off the rig before it comes to a stop, and they go to work in a blur. It's hard not to watch, not to be so impressed by the coordinated aggression of the scene that I forget to do my own job.

The truckies start working on the building's front door, a big roll-up made of corrugated metal. The first guy to the door reaches high above his head and buries a whirring circular saw in the metal, throwing sparks in every direction. He draws the saw down to the ground, then starts again from the top. When he finishes, another fireman gives the door a backward mule kick, and a large triangular cutout falls away into the smoky darkness, opening up a new hole to enter through. Ladders slam against the side of the building, and men run up the rungs and disappear over the lip of the roof parapet.

The first-in crew has an attack line in place, charged with water and ready to go, and they move inside as soon as the door is cut open.

They're going inside to find the seat of the fire, the root of the flames. Spraying water through the window won't do anything but keep the fire from spreading to the next building over. To actually put something out and keep a building from burning to the ground, you have to go inside.

There's flame from end to end, almost a solid city block of fire, and I don't even know where to start. It doesn't seem possible that we can get a handle on this thing before it burns to the ground.

My captain is standing by the side of the rig. His turnout coat is halfway unbuttoned, his boots are unzipped, and he's smoking. Smoking and talking calmly into the radio. He sees me standing in front of him and holds the mike away from his mouth for a second.

"What are you waiting for?" he asks. "Go fight some fire."

"Where?"

"Hell, I don't know. Anywhere probably. Pick a place." He points at the hose bed on the back of the engine, then makes a little snaking motion with his hand and points to another open doorway down the block that's blowing smoke. I grab a hose line and make for the door, stretching the heavy canvas line over the spaghetti of other hoses in the street. I can't believe he's sending me off alone, but there's fire everywhere, and the second-alarm engines won't be here for another five minutes at least. Long enough to lose the building if we don't do something.

"Help me! Help! Hold my baby!" A woman runs toward me with a tiny clump of blankets in her hand. Her eyebrows are singed and ashy. She's barefoot, and her T-shirt is on backward; she's naked from the waist down. She thrusts the baby toward me, but my filthy hands are filled with the hose and my ax. I'm torn. There are men inside taking a beating as they wait for me to hit the fire from the other end. I look at the baby's face and see it's covered with something black and tacky, maybe tar from the roof. The kid is screaming loud enough to momentarily drown out the noise from the sirens and the chain saws working on the roof. *At least it's breathing,* I think. The mother scans my eyes frantically, as if she senses I can't help her. She wheels and starts to run off, but I drop my stuff and grab her shoulder. She gives me the baby, and we run together to the corner, where an ambulance is on standby. We're both running in a crouch, as if an imaginary helicopter is spinning its rotors inches overhead. I dump the tiny baby in the center of the gurney, yell "You got it!" to the medics, and run back toward the fire without saying a word to the mother.

Smoke fills the air now, banking down almost to ground level. Streetlamps make lonely little islands of light in the gloom. It's midnight, but there are kids everywhere—kids on bikes, kids running back and forth along the sidewalk, kids flashing wide grins, apoplectic with delight to be witnessing such a miracle of destruction. They drift in and out of the smoke, just visible at the edges of my vision, their laughter and catcalls rebounding in the night.

"Whatcha doin'?" one kid asks as I stumble forward, my feet tangling in the slack hose.

"I'm working!" I shout, louder than I need to.

"You want some help?"

"I got it."

"I'm gonna be a fireman when I finish school. You guys get paid good, huh?"

"It's all right. I can't talk right now."

"Do you see a lot of dead people? I saw a dead body once. It had maggots and shit all up in its eyes." The kid jumps back onto his bike, loops a circle around me, does a little half jump over the charged hose line, and is gone.

My doorway isn't just smoky anymore. It's spewing flames now. There are other doorways on the row that aren't burning, but what would be the point of going into any of those? The awning above has burned through, and strips of melting plastic fall to the ground. The metal security door has been nearly ripped from its hinges, probably by our forcible-entry team, and the inner door stands open, flaming, waiting for me to go inside.

I lay the nozzle on the sidewalk and kneel in the broken glass to clamp my air mask over my face. The hose line is charged with high-pressure water, but I've got the nozzle turned off. The nozzle has a black plastic tip that is cracked and melted from previous fires. The handle is a pistol grip, and the chipped brass on/off lever is polished to perfection. Everything to this point has been done by rote, the same steps that I was taught in the drill tower: coat buckled, straps cinched, air bottle charged. I take my first breath of dry bottled air and look up at the flaming door. The fire is never going to go out unless somebody gets down on his belly and crawls under the heat to find the exact kernel of its origin. I should be exhilarated that that somebody is me. The heat radiating out of the doorway puckers the skin of my forehead in the thin, unprotected strip between the top of the mask and the brim of my helmet.

I reach for the nozzle but my hand comes up empty. I look around me in a panic; there's no greater sin on the fire ground than to be without a tool. *It was right there a second ago!* I stand up quickly to look around, but in my confusion I trip on my own feet and all of a sudden I'm on my back in the gutter. The belts and straps of my gear are tangled around me, and the heavy air bottle on my back leaves me flailing on the ground like an overturned turtle. The wind is knocked out of me, but my mind is racing.

"Get up, kid." Alvarez is standing over me with a smile, my nozzle held tightly in one hand while he pulls his mask over his face with the other. I'm not confused anymore.

Alvarez had snuck up from behind so he could steal the nozzle out from under me. It's part of the strange mind-set of firemen that they fight over the nozzle, sometimes even bowl each other over in the street in their zeal to be the first one through the door. Whoever gets his ass kicked hardest by the fire is the best firefighter. The man in front gets to make all the decisions, take the biggest risks, eat the blackest smoke. The guy on the nozzle is always the first to fall through a hole, the first to get burned, and, of course, the first to take credit for stopping the flames. I'm pissed at being beaten out of a job that I barely wanted to do in the first place. But I can't really blame Alvarez. He saw an opportunity and took it, like a boxer going in for the knockout when his opponent drops his hands. This is his job, and this is how he's always done it. Alvarez sticks out his hand and helps me to my feet.

"See you inside, kid."

My "partner"—a loose term for someone who just gleefully tricked me in order to steal my glory—extinguishes the doorframe and looks back to where I'm trying again to put on my mask. I can see him smiling through his own face piece. *Come on!* he motions with his hands before he disappears down the darkened hallway.

And then it's just the smoke again. It's too thick to see anything other than the smoldering doorframe directly ahead. But even with a crusty veteran leading the way, there's something in me that balks at going inside. There's no reason I should be going down that hall after him except that he expects me to. There's no rational explanation for why I'm about to voluntarily throw myself into the great flaming unknown, a place my every instinct tells me I should be fleeing.

The hose trails out in front of me, dragging forward slowly but constantly. Alvarez ran into the building with no more consideration than if he were getting on an elevator for the trip up to the office. I'm not there yet. I'm not sure I'll ever be.

When you don't know anything, when you haven't been to enough fires to put any faith in your own suspect talent, there's no choice but to give the moment over to trust. You trust that the man already inside is steadier and more skilled. You trust that whoever built the building did a solid job, that they hammered all the nails in up to their hilts, used sturdy lumber, and didn't cut any corners. You trust that your equipment is good and that all the guys around you will do their jobs well. You trust in the fact that firemen have always run into fires and that, far more often than not, they come out unscathed. And you trust finally in luck and in chance and in trust itself—you repeat with unexamined certainty the same thought that every single firefighter around you is having. You tell yourself that here, today, on this fire, in this place, it will not be your turn to fall. I put my hand down on the hose and follow Alvarez inside.

· 2 ·

The Tower

On the Saturday before the academy started, our new drill instructors made a show of hospitality and invited all the recruits and our families to the drill tower for an orientation. Everybody referred to the academy as "the tower" because of the four-story, blocky concrete building rising out of the center of the broken-asphalt yard. We stood around the barren yard nervously shaking hands. I did my best to approximate small talk, one of my least favorite activities.

I was there that day with Shona, a new girlfriend, who not only excelled at chitchat but somehow managed to be genuinely interested in every single person she met. I stayed close and let her be charming while I tried to deal with the sick feeling of nerves. I wish I could say I'd met Shona after I rescued her from a burning building, or that I'd defibrillated her back to life and, when she opened her eyes, she fell immediately in love with me. But instead I'd met her at a dinner party, two months before the academy started, before I'd even started telling people that I was thinking of being a firefighter.

After the initial bout of terror at making the first phone call to ask her out, I'd fallen for her more quickly than I could have imagined. I wondered—like every other person who's ever fallen in love—how somebody this good could have gone this long without already having been snapped up and lost to me. Shona stood barely as high as my chest, but she was a tightly wound rubber-band ball of determination and energy and exuberance, with just enough mischief and ridiculousness about her to make me realize that I'd hit the jackpot. It didn't feel strange at all that she was the person I chose to bring with me to the "friends and family" orientation after only a few months together.

Before the formal program started, she and I climbed the training tower and stood on the top platform looking out across the city. Oakland is not a place that is particularly attractive at first glance; few people come to visit unless they have business here. The city has become one of the busiest ports on the West Coast, and instead of beaches and scenic coastline, we have creosote-soaked docks and miles of rail yards. Instead of picture-book parks and landmark skyscrapers, we have windy warehouse districts and giant cranes for offloading cargo ships. The view from the Oakland highlands is memorable, but only because they overlook the hilly majesty of San Francisco.

Oakland is a city that feels like it exists on the cusp of a breakthrough, though what lies on the other side is altogether unclear. Always San Francisco's homely little sister, Oakland has been a mire of decay and neglect for decades. In the perennially mellow Bay Area—San Francisco had its Summer of Love, Marin County has its high-income crystal fondlers, and Berkeley has . . . well, Berkeley—Oakland has always been simply violent. Oakland was notorious for being on the leading edge of the crack epidemic and for being one of the main rallying points for the Hells Angels biker gang in the seventies and eighties. Even our football team, the Raiders, glories in a homicidal team image, and its war-painted fans pride themselves on their hooliganism.

During my childhood in Oakland, most of the city was off-limits.

There was no spoken prohibition; we just never went there. My experience of Oakland was my well-to-do north corner of town and a thin band of acceptability that ran along the line of the hills. The only time we came into the flats of Oakland was to go to the Coliseum to watch an A's game.

Now, standing at the top of the tower with Shona, it struck me that I'd never even been bothered by my utter lack of familiarity with the city I grew up in. The part of Oakland I knew was effectively the southern tip of Berkeley; the son of a doctor and an academic, I'd always faced directly toward the university instead of looking over my shoulder at the gritty industrial world that was the true Oakland.

"Are you nervous?" Shona asked me.

"No," I lied, zipping my coat down, then zipping it back again. "I just want to get this over with."

"Well, maybe you *should* be nervous."

"Maybe. But I'm not—is that okay with you?"

She stretched up on her tiptoes to kiss me, and I ducked away, afraid that one of my new classmates or drill sergeants would see me having a moment of human warmth. I expected the academy to be a nightmare of military discipline, and I was planning to survive for the next sixteen weeks by allowing myself all the emotional range of a terra-cotta warrior. I knew I wouldn't be the strongest trainee, or the most competent. All I wanted was to be the blandest, the one who never got singled out for anything. Being publicly kissed during orientation would be a disastrous beginning for the newly minted man of steel.

Instead Shona put her hands over mine to stop my fidgeting with my coat. "You are such a freak," she said, laughing. "You know you're going to be fine, right?"

I didn't answer.

"Well, at least they all look like nice people." I wondered which people she had been looking at.

"Come on, we have to go," I said, looking at my watch.

"We don't have to be down for another ten minutes," Shona said, wheedling, teasing. "I like it up here."

"I know, but it'll take us two minutes to get down the tower, and then I have to use the bathroom first, which will probably take three minutes, and I want to make sure I get a decent seat close to the front so I can hear, but not right in the first row. And I want to be there at least five minutes early to make a good impression. So we're pretty much late already. Can't we just go?"

Shona eyed me with suspicion. Or pity. "You're not going to be any fun at all for the next four months, are you?"

"Look, I can't worry about that. We're *late*." I zipped up my jacket, unzipped it, and ran down the stairs.

I heard the fire trucks before I saw them. Not sirens, just deep, throaty diesel engines rumbling down the long lane toward the drill-yard gates. The drill instructors had told us to assemble with our families in one corner of the yard and stand out of the way for a demonstration. The fire trucks pulled up in front of the concrete tower and disgorged fire-fighters from every door. They went to work quickly, pointedly ignoring the new recruits gaping at the spectacle. I heard warning alarms, the crackle of radios, shouted directions, and the slamming of doors. Two firefighters dragged heavy hoses from the back of one rig; the massive aerial ladder telescoped outward in increasingly narrow sections. Some of my fellow recruits leaned over to their wives, explaining the proceedings.

"Do you know what it is they're trying to *do?*" Shona whispered in my ear.

"I have no idea," I answered.

On the ground there was hose everywhere. Somebody hoisted a chain saw, and other people charged up the fire escape calling out unintelligible commands. One firefighter looped a length of hose over the railing of the third floor and tied it off with a bright yellow piece of rope. Clearly they were showing us something dramatic, but it felt like being in a foreign country watching the natives perform some mysterious ritual.

When a spray of water came bursting through the window of the fourth floor, everybody cheered. Shona and I joined in; it was the thing to do. The firefighters on the ground took off their helmets and gloves and broke into smiles, congratulating each other for achieving the great something-or-other they'd just finished. Whatever it was they'd been doing was an obvious success, and I felt an unexpected swell of pride, as if I were the one who'd just accomplished something. It was the same glory-by-association feeling I used to get as a kid when Rickey Henderson would steal a base for the A's or Sleepy Floyd would hit a sweet jumper from deep for the Golden State Warriors. I wasn't on the field yet, but these guys were clearly playing on my team. A few of the firefighters went over to the new recruits where we stood in our clean civilian clothes. They sought out the friends, cousins, sons among us and wrapped them up in hearty handshakes and manly backslapping embraces.

"Let's go! Let's go, recruits! Everybody into the classroom now! We're *late*," a man in a perfectly pressed uniform yelled across the yard, and we all started running. I left Shona behind to fend for herself and sprinted for the front. The academy wasn't officially open until Monday, but there we were, running, everybody afraid to be the last one inside.

Inside the corrugated-metal garage that doubled as a meeting room that day, the instructors—they called themselves "the cadre"—separated us from our families and sent the wives and kids off to spend a few minutes with a box of doughnuts and a giant urn of stale coffee. Those of us who remained sat still in uncomfortable folding chairs, chattering nervously. One of the men in uniform raised his hands for us to be quiet and stepped to the front of the room. He looked doughy and pale and had the sad buzz cut of a balding man trying to appear rigorous and hard.

"I," said the fat man, "am a runaholic. There is nothing I like better than running. You should expect to run. Every time you make a mistake, you will run. And everyone will make mistakes. I can run all day, and I love it." He spoke as if he'd been practicing, as if he'd been trying out the speech in front of his mirror for days. The fat man was wearing his Class A dress uniform, all brass and bugles, polished to perfection. A belt buckle was occasionally visible when his belly bounced up and away from it. He'd clearly bought the jacket years earlier, during a skinnier phase of his career.

"My name is Lieutenant Burton, and for the next sixteen weeks you belong to me and the rest of the training cadre. We'll work you hard all day, and when you go home at night, you'll be studying until it's time to come back the next morning. Tell your wives and kids not to expect much out of you; this is going to be hard on them. But it will be even harder on you."

I looked around for somebody to share a smile with. Burton's speech was such pat Hollywood you're-in-the-army-now drivel that I felt we were being set up for a joke. More than anything, I felt embarrassed for him. Could he possibly be serious? But everyone around me was rapt. For the first time in my life, I was self-conscious about slouching in my chair. I uncrossed my legs and sat up straight, stopped bouncing my feet and playing with my hands. *He's not kidding,* I thought. *He means this stuff.* The guy on my left looked as if he'd been turned to stone.

As Burton droned on about how we would run until our feet became bloody stumps, I couldn't help but notice the piece of doughnut dangling from the corner of his mouth.

I'd been expecting Patton; instead I would be at the mercy of Elmer Fudd.

. . .

I'd been admitted to the academy late. Through a clerical error, I'd found out only on Friday that I would be starting with the fire department the following Monday. After the orientation I spent the rest of Saturday scrambling around Oakland to buy the pieces of my uniform: a certain specific brand of black work pants from a store in East Oakland, a black leather belt from a police-supply shop, taken from the rack next to the baton holders and the body armor. I had tried to act nonchalant as seasoned firefighters and cops milled around, joking roughly with each other and picking out gear. Afterward about five of us new recruits arrived simultaneously to pick up our regulation T-shirts at Station Eight, the firehouse I'd visited on school trips as a kid. The firefighter in charge of the union haberdashery was a round little man with no facial expression and a carnival barker's handlebar mustache. "Buy a big one and a small one," he said, passing out fire-department sweatshirts. "You'll get skinny as hell while you're in the tower and then end up fat like me as soon as you get out."

Since I had been added to the roster late, there were no boots waiting for me at the uniform store. On Sunday night I reached under the bed for my sturdy leather hiking boots. I leaned out the window of my house and smacked them together to knock off the red dirt that I'd brought home with me from my last job as a seasonal park ranger in Utah. Standing in front of the mirror, I looked ridiculous: crisp new T-shirt, pressed black pants, and scuffed red boots with unevenly worn soles.

The next morning I could feel everyone's eyes on me. They all had shiny black boots with laces on the side and zippers up the middle. I spent the day standing behind people, but I turned as red as my boots every time somebody looked me up and down and then shook his head when he got to my feet.

"You been traveling, son?" One of the drill instructors snuck up behind me and laid a hand on my shoulder. His voice was silky smooth and malicious. He wasn't making small talk.

"Uh, no sir." I felt self-conscious; I'd never called anybody "sir" in my life, and I was sure I would say it just wrong enough to be exposed as a fraud. I'd grown up on the edge of Berkeley, and all my friends' parents had been named Don or Jim—never "sir," not even "mister." I felt silly saying the word, but it seemed like the thing to do in the tower.

"So why are you wearing hitchhiking boots? This ain't no trail." Captain Gold was slim and handsome. I recognized him from the orientation as the man whose bark had sent us running. He stood militarily erect, seeming tall without actually being tall, and he looked like he'd been born wearing a uniform. When he spoke, he was slow and deliberate, and he gave off the impression that there was nothing he did not see. ("Gold?" my mom would say later, hopefully. "Is he Jewish?" "I don't think so, Mom, unless he's a Sammy Davis Jew.")

"I'm sorry, sir. I got my acceptance letter late and—"

"Now, why would you think I care? I don't want to know *why*. I just want you to look right. Everybody else here looks right. Do you think you're special?"

"No sir, I just—"

"Well, good. Because I don't think you're special either. Right now you're the only person here who isn't right, so I'll be *watching* you special—and that's the only kind of special you're going to get. Don't make me see those traveling shoes again tomorrow. Is that clear?"

"Yes sir, very clear." I'd never had a first day like this. Where was the name game? Why weren't we sitting around in a circle talking about our fears and aspirations for the upcoming program? I was shaky and sweating; it wasn't even lunchtime, and I had already pissed off the scariest instructor. The orientation with Lieutenant Burton had lulled me into thinking that the instructors would be harmless buffoons. Clearly, Gold was a ramrod-straight force of nature, a pillar of terror to be reckoned with.

"You got your first beat-down, didn't you?" A man about my age with alert eyes and the first smile I'd seen all day came toward me and offered his hand. He was short and a little thick around the middle, but what struck me first was how totally at ease he seemed, as if he hadn't noticed that he was a lower-than-whale-shit recruit like all the rest of us.

"It's nice to get your spankings out of the way early," he said. "I'm Antoine. Nice to meet you." I took his hand, glad to have found someone who seemed like a normal human being. "It's always the same," he said. "Good cop, bad cop. Burton's a clown. Looks like Gold's going to be the heavy for this tower. I was a firefighter in Modesto for seven years. I used to *teach* these academies, so I know all their little tricks." Antoine carried himself like someone who knew that he would get the best of any situation he found himself in. It wasn't the swagger of a big man, but the coolness of someone who had complete confidence in his abilities. He rarely stopped smiling, but I could tell that he took the academy very seriously. He was like Gold: He saw everything and seemed to be filing away information to be used later.

"Thanks," I said, letting out my breath finally. "I just couldn't get boots in time. I wasn't trying to be disrespectful."

"He knows that," Antoine said. "He just needed something to ding you for. It's his way of getting the game started."

"I hate to piss that guy off on the first day."

"You didn't. He's setting the rules. He's already forgotten about you."

Antoine's reassurances were nice, but, glancing around, I realized that Gold was right about one thing: I looked nothing like a firefighter. In my family there were no uncles, brothers, or even neighbors who had been firefighters. All around me in the tower were men with short hair and square jaws, guys who paid attention to the details of their appearance. I could imagine every one of them becoming firefighters, but I had trouble conjuring a mental picture of myself leaning against a fire engine and talking in an assured voice about "that ass kicker we had in a high-rise over on Jefferson Street." There's just something about me that resists a sharp, crisp look. I'm the guy who always has his shirt half untucked in the back, the one with spinach stuck in his teeth. I'm the guy whose eyebrows meet in the middle. I have no ass whatsoever—my legs lead straight into my back without interruption. My feet stick out to the sides, and I walk with a duck-footed, assless little shuffle that is decidedly antiheroic.

"But he *is* right about one thing," Antoine said. "Those boots have got to go. Check with me later, and I'll tell you where to go to get right."

Though fire fighting can't be learned by doing anything other than fighting fire, an attempt at education must be made. Oakland's drill yard was in a shoddy industrial neighborhood halfway between the bright lights and trendy shops of Jack London Square and the littered train tracks and dark warehouses of the docks. The rear fence of the yard butted up against a fetid slough where murky water rushed back and forth from Lake Merritt to the estuary. In the late afternoons, the smells of algae and mud would compete with the choking aroma of burned coffee beans coming from the roasting company on the other side of the yard.

The buildings of the Training Division were all old trailers, prefab metal buildings that managed to look both temporary and run-down at the same time. A homeless woman lived just outside the entrance; she'd squatted in an abandoned shack and built up an impressive little city of two-by-fours and blue plastic tarpaulins. She had dogs and a small garden, and she seemed more permanent than anything inside the drill yard. In the mornings I would park on the street outside the gates and walk in. "Don't work too hard now," she'd call out. "It don't get you anywhere."

The drill yard itself bore an uncomfortable resemblance to her homeless minicamp. It was ringed by barbed-wire fences and looked like the aftereffects of a tornado in a junkyard. Stationed (or more likely dumped) around the edges of the yard were black-and-white police cars stripped naked of their light bars, a van with a jagged hole where a roof should have been, a light airplane with no wings listing to one side. Fire hydrants dotted the periphery, some of which would flow water when the spindles were spun, some simply mounted on plates to be used during dry runs.

In one corner of the lot was the forty-foot-tall front of a building, propped up tenuously by timbers from behind, like a façade from a Wild West movie set. The wall had windows, a parapet, and a heavy spray of pockmarks from a hundred thousand ladders that had been slammed into place over years of training. At the base of a wall was a door, barred from the backside with wooden slats that would splinter when forced, so recruits could get the feel of breaking open a door without having to rebuild it between attempts. Alongside the roofless wall was a wall-less roof, a sharply pitched shingled slab perched atop four thick, round posts. An ancient railcar hid in the weeds alongside a rusted cube truck, a mailbox, and an immense, soot-blackened fuel tank.

But the defining feature of the drill yard was the tower. It looked like a Soviet-bloc high-rise, six stories of chipped concrete with empty holes for windows. The exterior stairway was fenced by the same dead-gray metal railings as the fire escape on the opposite side of the building. Heavy storm doors were sunk into the ground, and the basement air around them smelled like mildew and ash. The structure stood like an abandoned guard tower, keeping watch over an apocalyptic parking lot. When it was first built, the lot had been on level ground, but as the estuary undercut the land, the area subsided. The tower, resting on pilings buried deep in the muck, never sank an inch, only appearing to rise as its world gave way around it.

For sixteen weeks that tower was the focus of our lives. I'd been watching it for years; it sits below and to the side of a freeway overpass, and whenever I passed, I would crane my neck to look down until the road turned away. In brief seconds through the windshield, I had seen men laboring up the exterior stairway laden with heavy coats and air bottles, holstered axes slapping at their thighs as they laid hose, circling upward from one landing to the next. Once I became a recruit, the tower immediately lost any glamour it might have had from a car window at sixty miles per hour. Every day we climbed it, rappelled off it, broke down its doors. We dragged hose up from the bottom and threw ladders to the top. We ran laps up the staircase and shimmied down the fire escape. "The tower" became the code for the entire experience of training. We'd say, "Only six more weeks until we get out of the tower" or "You can't let the tower get you down" or "The tower is killing me."

My recruit class (named "Three-Nine-Eight," a call sign that we were required to holler maniacally while doing most things) was made up of twenty-one men and three women. Gender aside, we were a group that resembled the city we were about to serve: black, white, Hispanic, Vietnamese, and Filipino recruits, lined up side by side. "Together, all of you add up to nothing," said Lieutenant Burton one day. "You are all the same to me, because none of you know anything about fighting fire." I wondered if Burton ever said anything that he hadn't rehearsed beforehand.

To a certain extent, he was right, but I knew even less than most. Like Antoine, several of my classmates had left suburban departments for the promise of big fire and greater opportunities in a large city. They'd had to start all over again at the bottom to "learn the Oakland way," but the loss of seniority was balanced out by the advantages of working for a gritty, urban department.

Many of my classmates were the children or brothers of firefighters. They talked about how things would *really* be once we got out of the tower, and they lamented the changes in the fire service since the days when their fathers had had the run of the city. Most of the recruits had ridden along as volunteers with fire departments (many in Oakland), and almost all had at least a few college courses in the rudiments of emergency medicine and fire science.

"How many tests did you take, Unger?" asked my classmate Dwayne on the first day. He was impossibly tall and gangly, with crooked teeth and a head so large that the instructors had needed to special-order a helmet for him. He walked with a foot-dragging sort of lurch and was something of a savant when it came to recitations of the minutiae of fire-department life.

"I think I took all of them," I said. "The written, the psych, the physical agility—"

He cut me off. "No, I mean, how many cities did you test for?"

"Just this one. I saw a job announcement on a bus bench."

Dwayne looked at me as if I were crazy. "You might not want to tell people that, Unger. I tested for twenty different cities before I got this job. San Francisco, Stockton, Los Angeles . . ." I quickly found out that Dwayne was not unusual. Many of my classmates had been methodically preparing for a life in the fire service from the moment they graduated high school. They'd studied, tested exhaustively, and kicked around small volunteer departments like minor-leaguers. They all considered Oakland to be their call-up to the Big Time, the pinnacle of fire-department success. Dwayne wasn't embarrassed by having taken such a long road to get where he was. Instead I learned that my classmates viewed their years of trying to get hired as a badge of honor. The desire to fight fire burned in them; I was expected to be ashamed of my own whimsical ambitions. For me, the fire department had started out as a lark. Dwayne just shook his head. "You don't even know how lucky you are."

Historically, fire departments have been filled by the children of firefighters and by tradesmen looking for more excitement and a steadier paycheck. Knowledge of how buildings come together is very valuable when they start to come apart. A firefighter is a contractor's evil twin brother, always tearing down, ripping up, making things ugly. Our class had a plumber, a drywaller, a mechanic, and a carpenter. Even those who were not specifically from the trades tended to be skilled workers, people who would die with a hammer in their hand before they called a contractor to work on their own homes.

The fire department also attracts plenty of former military men. For them, fire fighting has a comfortable paramilitary structure with a heroic common mission, an identifiable chain of command, and a familiar tradition of self-sacrifice. Russell Earle was one of our soldiers, an air force reservist who had flown in Somalia and Bosnia. He'd been a navigator for a large assault plane, and it was clear that he reveled in crisp clothing and spit-shined boots. He had a high, nasally voice and nervous, fervent eyes that he hid behind thick glasses. Every morning during the academy, as we lounged around before roll call, he'd yell out, "One minute, guys! Come on, let's clean it up! One minute till inspection." It didn't take long for everyone to mimic him, pinching their noses and whining a chorus of "One minute, guys." Russell was well intentioned but *tight*, always on the lookout for the five-star general who would never appear. His exasperation was harmless, so the guys teased him gently, gave him nicknames until "Russell Earle" was entirely replaced by Rusty Girl, Rear Admiral Buttercup, and Russell the Love Muscle.

After the firefighters' kids, the tradesmen, the military folks, the former firefighters, and those who had come from other civil-service jobs like customs or the probation department, there were only a few complete outsiders left. "What did you do before you got hired?" was always the first question following introductions. I tended to choose the most macho of my recent short-term jobs. "I worked at a mine site," I'd say, never mentioning my stint driving the muffin-delivery truck or the months I'd spent in Utah monitoring peregrine falcons' mating behavior. Mostly I just stayed quiet, tried to downplay my existence. I figured if I could just make it through the academy without being called out as a fraud, I'd end up a firefighter like the rest of them.

Mary Mills was another total newcomer, though she was tall and well muscled from her previous life as a personal trainer. She was looking for more of a challenge than just putting people through their paces at the gym. Mary had heard about how hard the job could be on women, but she was determined to do it successfully and to prove a point at the same time. Though we'd never met before the academy, I recognized her immediately: She was Berkeley to the core. Her constant hangdog expression told me that she hadn't yet made her peace with the fact that being a firefighter recruit meant that she'd spend her new life being screamed at.

I'd grown up surrounded by people like Mary Mills. I'd always attended earnest, progressive day schools where everyone was made to feel cuddly and special. At recess we ate teeth-shattering banana chips and drank cloudy apple juice before returning to our cooperative, non-violent games. We sang, we shared, we had "circle time." Our teachers asked, "How did everybody *feel* when Jenny broke her shoelace?" We expressed our emotions. I could see immediately that Mary Mills was cut from the same cloth, and though I tried to hide, she saw it in me as well and latched on. From the first day, she carried with her a look that was midway between fear and resignation, like a sheep surrounded by wolves. I didn't want her to feel close to me. I could tell that the tower was all about endurance and survival and locking yourself away somewhere until you got your badge. Nobody much cared if you felt taken care of or not. I had to watch out for myself. Too much love and emotion wouldn't help me get to where I needed to be.

It didn't take long for us to fall into a routine. Since we'd been told that a single instance of tardiness could result in dismissal from the academy, our cars were often lined up outside the locked gates an hour before any of the instructors arrived. "You need to leave home early enough," Gold said, "so that if your car breaks down you still have time to *walk* here. Don't even try to give me excuses." In the predawn hours, we'd huddle together comparing notes or try to catch a few more desperate minutes of sleep on car seats at full recline.

In a profession as given to chaos as fire fighting, I was amazed at how much of our time was spent memorizing minuscule details. Every night I went home and added to my stacks of flash cards. A lifetime of valuable baseball trivia had to move aside to make way for the zip code of every Oakland firehouse, the number of cutting teeth on each of the different brands of chain saw, the specifications for the different types of hydrants we might encounter. We had six binders full of paperwork that hadn't been updated since the 1970s, and every detail was fair game for weekly tests. Dwayne, with his appalling love of details, would raise his hand and ask, "Do you want us to memorize the product booklet that came with our new turnout boots?"

"Well, what a good idea, recruit. I was just about to assign that myself," Burton would answer to a chorus of groans, wiping the remnants of lunch from his face with the back of a fleshy pink hand. "Make sure you memorize the serial numbers, too," he'd say.

Recruits don't sleep much. In the evenings I dragged myself home, tried to rehydrate, and fought against sleep so I could get in a few more hours of studying. It was not a good time to be starting a relationship, and I wondered what Shona must think of dating a man who wore nothing but black jackboots and Ben Davis work pants.

"Can you come out tonight?" she'd ask. "I want you to meet some of my friends."

"No," I'd say. "I have to memorize the City of Oakland mission statement and the rush-hour capacity for all the different kinds of subway cars."

"Can't that wait?"

"No, it's really important." I knew how ridiculous I must sound, and

I just hoped she'd stick around long enough the meet the Zac who didn't devote every second to rote memorization. When I'd finally gotten up the courage to tell my grandfather that I was quitting graduate school to be a firefighter, his first comment was, "And the kind of wife you want is going to be happy with marrying a fireman?" I could only cross my fingers and hope so.

In class I took copious notes and pretended that I understood what everyone was talking about. Like most people in the world—civilians, as I learned to refer to them derisively—I had used the terms "fire truck" and "fire engine" interchangeably; in the fire department, this is as grievous a sin as confusing credits and debits would be to an accountant. A fire engine, I learned, is the smaller and more common of the two types of equipment. Fire engines carry hundreds of gallons of water and thousands of feet of hose. It is the job of the engine to attach to the hydrant for water supply; firefighters on the engine are the ones who will actually pull attack hose from the rig and make entry into a burning building in order to extinguish the fire.

A fire truck—also known as the "aerial" or by civilians as the "hook and ladder"—by contrast is the long piece of apparatus, the one with the firefighter sitting in back to pilot the rear wheels. Trucks are like giant rolling toolboxes, the cabinets filled with every imaginable implement of destruction, from chain saws to pickaxes to the Jaws of Life. Trucks, unlike engines, carry no water. Their main cargo is ladders, ladders, and more ladders, including the massive mechanized aerial ladder that extends from on top of the rig with a hydraulic whine.

I suppose I'd always known that firefighters used ladders, but I'd never given any thought to why. Every fire I'd ever encountered—campfires mostly—had been conquered solely through a liberal use of water. House fires, however, must be vented, the superheated gases allowed to escape into the air. It is for this reason that firefighters throw ladders and climb onto roofs—in order to cut holes to ventilate the fire. During the classroom course on ventilation, I played along as if I had always known about this aspect of fire fighting, when in reality it was a wholly new and terrifying part of the job.

Engines need trucks and trucks need engines. No single piece of apparatus can fully extinguish a fire. We learned that the two must work in concert. The truck crew forces open the door through which the engine crew will bring their hoses. The engine crew puts water on the fire, the toxic steam and gases of which must be dissipated through the holes that the truck crew has cut in the roof. Truck work tends to require more heavy lifting and is statistically more dangerous than engine work. Because of that and because there are far fewer trucks than there are engines, being a "truckie" takes on a certain mystique. Truckies tend to be big and, in the view of "hose men," dumb. By the third or fourth week of the academy, recruits were already starting to decide if they were better suited to truck or to engine work.

Roll call came every morning at 7:55. In any weather we'd gather listlessly behind the trailers until Russell issued his useful but much-maligned "One minute, guys." As we lined up in groups of six, Gold and Burton would prowl down our aisles inspecting our outfits. A torn T-shirt or a faint shadow of beard stubble was cause for a written reprimand and the ever-popular threat of dismissal.

As far as my boots were concerned, I couldn't even approximate success. I'd never worn anything but tennis shoes in my life, so the entire idea of shining boots was mysterious and unpleasant. I went to a drugstore and bought every single shoe-care gimmick sold. I must have misused my arsenal, because my boots were always a dull mockery of the retina-shattering black that all my classmates managed to achieve. The truth was that, no matter how many disapproving looks Gold gave me, I simply couldn't force myself to care about this one detail. At night, after studying, all I wanted to do was sleep; I couldn't bear the prospect of spending forty-five minutes buffing boots that would be scuffed three minutes after roll call. Every morning I regretted the previous night's laziness. As Gold stalked toward me, I'd resolve to do better, to care more about Johnson's Ever Varnish Black and Turtle Magic pads.

After the recitation of the Pledge of Allegiance—an American standard that I had never had occasion to learn in my socialist-inflected grade schools—we broke for housework. This was invariably an hour spent dodging Gold and trying to get in some extra studying. In addition to our twenty-four recruits, there were two other classes of trainees in the tower at the same time. Seventy-five man-hours a day was more than enough to keep the trailer park tidy. So, mops in hand, we'd hide in the bathrooms or pretend to pick weeds at the far corner of the lot.

One day I surprised Russell Earle in a broom closet. He twitched visibly when I opened the door, but after seeing it was only me, he went back to his lonely monologue. "The Tempest 2710 Blower weighs 79 pounds and blows 4070 cubic feet per minute. The Honda generator has 20 amps and 125 volts . . ."

Mornings and afternoons were spent on the yard, our time divided between hoses and ladders. Everything was new to me, and each day built upon the one before, until my mind was full of the different variations of hose lays and ladder throws: schoolyard return to pump, offside scaler, auditorium raise, two leads into a standpipe. We learned hose lays for a dozen scenarios, like low-pressure hydrants or long alleys requiring huge extensions of hose line.

We learned that no matter what type of hose lay is called for, no matter whether the engine begins at the hydrant and lays hose to the fire or begins at the fire and drives back to the hydrant, the objective is always the same: "Put the wet stuff on the red stuff." The only way to do this is by ending up with the nozzle at the front door. And since the bottom of every nozzle is a female fitting, the last length of the hose must either be naturally male or be made to function as a male through the use of an adapter. "It's simple," said Burton, savoring the upcoming line. "Water has to come out, and it has to come out from the male end. Dick to the fire. Just remember that. Dick to the fire and you'll always be okay."

At the end of the day, after we were all battered and broken from hours of hoses and ladders, Gold would gather us up for physical training— PT. On alternate days we would run a three-mile course around Lake Merritt or work through a circuit of weights and exercises. One Friday, Gold called us out to do the "STAR" drill, an acronym whose meaning would remain obscure to us. The STAR drill was similar to the physical-agility test that we had taken as applicants. It was an obstacle course that replicated the demands a working firefighter faces on the job. Wearing a heavy coat and breathing air from a tank, each of us would pull wet hose, drag a weighted dummy, climb the tower, and hoist a weight up through the window. I had been working out hard in prep- aration for the academy, running in the hills and climbing stairs at the stadium with a heavy backpack. I'd never felt more fit, and I was deter- mined to have the best time of the group. When Gold hit the stop- watch, I went out at a dead run.

Through the ladder hoist and the hose drag, I was feeling strong. It was the first time I had ever breathed air off a fire bottle, and I liked the sensation, the focus, the narrowing of vision to the task at hand, the re- assuringly loud hiss of an air intake that never lets you forget where each breath comes from. I sprinted around the cone at the far end of the yard, then hand-over-handed a hundred feet of dead weight hose until it was flaked neatly behind me.

At the foot of the tower, I snatched up the hose roll and flung it onto my shoulder. At the balcony in front of the second floor, I could feel my chest starting to tighten a bit, but I figured it would pass, so I took the steps two at a time to finish quicker. On the fourth floor, my lungs burned and sweat rolled off my forehead. I dropped the hose roll on the floor and swatted ineffectively at my masked face to try to clear my eyes. I was taking huge gasps of air, demanding more than the regulator would provide. Looking out the window, I saw Gold on the ground, stopwatch in hand, standing next to a weight on the end of a rope. He was screaming up at me, but all I could hear were the grunts of the air hose and my heart pounding in my face. I took the rope in my hand and pulled—once, twice, three long armfuls of rope before I staggered backward. The room was shaking, and I sank to a knee. I was back behind the windowsill, down below so Gold couldn't see me, sitting flat on my ass, the rope in my hands with the weight dangling in midair twenty feet below me. I lumbered to a stand and leaned against the low wall. Grabbing the lip of the windowsill, I hung out over the edge, bending from my waist, watching the weight spinning on the end of the rope and the world spinning in the opposite direction around it. My only thought was that I'd never be a fireman. I didn't have the lungs.

I forced myself to finish the pull, hurled the weight up onto the sill while my brain was screaming, *This is what it's like to drown!* It seemed so easy—all I had to do was rip the mask off and I could have all the air I wanted. My hands were shaking, I was losing my grip on the rope, and I wanted to crawl out of my own skin and lie on the ground sucking in great mouthfuls of air. I don't remember letting the weight slide back down again; I was in such a panic to pull off my mask that I probably just let it drop.

As soon as it was gone, I fumbled with the chin strap of my helmet and ripped the mask off. The lack of air had made me worthless. My hands and feet tingled, and I dry-retched over a steel barrel of half-burned shingles. But the air felt so good, the fetid roasted-coffee breeze so sweet.

It didn't take long for relief to give way to embarrassment. I put the mask back on and hoped Gold hadn't seen anything.

"Piss poor, Unger," he said. "If that was a working fire, you'd be dead."

"Yes sir. I'm sorry, sir. It won't happen again," I said, thinking that it probably would.

"You're damn right it won't. Maybe you should just walk on out of here. Won't nobody say a thing if you leave your books on the table and never come back. Think on it, Unger. Nobody wants to keep you here if you don't want to stay." Mary Mills stood behind him nodding sympathy and flashing a big "stick with it" thumbs-up. Her overfriendly support just redoubled my shame, and I shifted around so I wouldn't have to look at her.

Gold had a point, though. There was no way I'd be able to make it through the academy, much less an actual incident. The STAR drill was a test, and I'd failed it. I knew that I wouldn't quit; I'd seen too many boot-camp movies to mistake Gold's threats for anything but boilerplate intimidation. But I was disappointed in myself. More than that, I was disgusted. When things got tough, my response was to panic, and I wondered why I had ever deluded myself into thinking that I might have some inborn gift for this job. If I'd taken off my mask in a fire, the heat and gases would have killed me, and yet as soon as I felt low on air, my first reaction was to take off my safety gear rather than try to calm myself and slow my breathing. I had thought that I would ace the thing, be the star of the STAR drill, but instead all I'd succeeded in doing was making a fool of myself. My classmates were right: You don't decide to be a fireman because you see an ad on a bus stop and think it might be a fun thing to try for a while. Everything I'd thought about myself—that I was smooth, in shape, cool in a tough spot—had been called out as a lie by one run up a shitty concrete stairway that didn't even have a fire in it.

Throwing the Fifty

At the end of winter, we were hit by El Niño, the once-a-decade cycle of ocean currents that brings torrential weather onshore to the Bay Area. Rain fell in constant dreary sheets, gutters backed up, and we spiked the feet of our practice ladders through several feet of standing water. The railroad ties that demarcated the parking lot lifted off and floated past us as we ran through our drills. Every day we would shrug ourselves into turnout coats still soaked from the day before and run furiously around the drill tower trying to stay warm.

Throughout, Medical Thursdays were a welcome reprieve. On medical days we got to sit inside, nursing our wounds as we listened to lectures and practiced patient-care skills. I sat next to Phan Nguyen, a Vietnamese immigrant who had jokingly given himself the nickname "24," a reference to the fact that he didn't care if he was the last-ranked person in our class, just as long as he managed to graduate. Though in front of the instructors Phan was all wide-eyed obedience, he was bright and wily, and he refused to be cowed by the inanity of the tower. He taught us enough Vietnamese to be able to say "Nice tits! One, two, three! You have a big ass! More chicken, please!" which we would scream while dragging hose, much to Gold's angry bafflement.

Phan was my partner the day we practiced dealing with emergency childbirth. The rubber mannequin we were using was stumpy and more than a little unsettling. Designed for teaching only this one skill, it was pared to the bare essentials: It cut off sharply above the belly button and below the upper thighs, a foot and a half of graphic groin complete with molded red plastic pubic hair. The instructor could bend down where the torso should have been and shove through an emaciated plastic newborn in various positions of distress. The setup came with a coiled rubber umbilical cord—with handy snaps for easy fetal attachment—plus an odd round placenta that looked like nothing so much as an oversize plate of huevos rancheros. At one point Phan forgot the proper technique for cutting the cord and was ordered to run three laps to the top of the drill tower. With the placenta on his head. Even Captain Gold grinned as we watched Phan in his clunky steel-toed boots trying to balance the rubber placenta while he charged up the stairway shouting, "I *will* wait until the cord stops pulsating before I cut it! I *will* wait until the cord stops pulsating before I cut it!"

Laughs never lasted for long in the drill tower, though. Gold was thundering again. "I am *not* happy!" he screamed. I'd been spacing out and missed the beginning of his speech. I didn't know what was making him unhappy this time, but I'd heard this part before. We were all standing at attention, lined up in four rows of six. Gold was dressed impeccably, as always, even in the driving rain. It was coming down in freezing streams, but he seemed not to notice. Hatless and full of fury, he stood erect and looked straight forward, with none of the hunching or ducking that comes unconsciously in a downpour.

"Some of you people are on the bubble! There is no reason I can think of why any of you should feel like you've got a right to your jobs yet. Don't you understand that? This ain't no country club. This ain't no junior high where you can just get by on your good looks. I'm cutting paperwork *today*—do you understand what that means? It means some of you won't *be* here tomorrow." He stopped for a minute to give himself a chance to fix his eyes on a few specific people in turn. "I've been coddling you people for far too long. All I've been seeing out of you is a lot of backslapping and happy-facing, and I don't understand what you think is so goddamned fun about this. It's week ten, and some of you don't even know how to hook a hydrant! Some of you don't know the top of the ladder from the bottom. And you expect to be firefighters? I don't think so. Things are going to change around here. Right now I've got my fuck-you boots on, and I'll shove them right up your asses if things don't get better. It's week *ten*, people. I expected more out of you."

I tried to picture what a "fuck-you" boot might look like: some sort of collaboration between Doc Martens and the Good Vibrations catalog, perhaps. Mary Mills was standing in front of me in the lineup, visibly trembling.

Gold paused and softened his voice maliciously. "I'm a happy guy. I like to be happy. And right now I am *not* happy."

Antoine leaned over toward me and whispered, "And I do *not* care."

As we broke out of roll call and shuffled off to our housework, Gold pointed a long finger at me. "Meet me in my office." Singular attention was never a good thing. Avoiding Gold's office had been the crux of my entire tower-survival strategy. I would joke around with the other guys but go entirely silent whenever a drill instructor neared. I worked hard, never asked questions or claimed to know anything. When the instructors contradicted themselves, I always apologized for my error so that they could save face. I'd bite my tongue and run laps up the tower without complaint as punishment for answering "every five seconds" instead of "twelve times a minute" when asked how often oxygen should be squeezed into an unconscious heart-attack victim. I had created a new, boring, humorless persona that I could slip on like a mask whenever I sensed a cadre member lurking nearby. Several months into the academy, a number of the instructors still couldn't be counted on to remember my name.

I stood in front of Gold's office for twenty minutes, waiting for him to make his appearance. I stood up straight, held my hands clasped behind my back, and affected an obedient air. I was scared and pissed. I couldn't imagine what I might have done wrong. Russell Earle came past me at an efficient trot with a mop in his hand. When he saw where I was standing, his eyes widened. "What did you do?" he asked. He'd been flying the straight and narrow, too, and the thought that I might have slipped up made him feel even more vulnerable than usual. "I don't know," I said. "Leave me alone." Admiral Buttercup shook his head and ran off, determined, no doubt, to do a better job than ever on the bathroom floors.

Gold finally arrived and walked past me without speaking. After a minute he called me in. "Sit," he said, pointing at a chair opposite him. "I said sit *down*, Sullivan." "Sullivan" was the default he called us by when he couldn't quite attach a name to a face. It seemed unwise to correct him just then. He leaned back in his chair, uncharacteristically casual, with his fingertips pressed together and tented in front of his face. He stared at me for a long time without speaking.

"So what's the problem here?" he said finally.

"Sir . . . ?"

"Is there something wrong at home?"

"No sir." I had no idea what he was talking about, and I was even more perplexed that he might actually care.

"You don't seem happy. And"—he broke into a wide, disingenuous smile—"you know how I want you to be happy."

"I'm sorry, sir. I really don't know what you're talking about. Everything at home is great. I'm just trying to keep my head down and get through this academy. Did I do something wrong?"

"No. But you're too quiet; I can't see your spirit. I don't want you to go postal on us one day. I want you to be one of the guys." He flashed his gleaming teeth at me again.

I was shocked. My stealth firefighter act had worked so well that I'd convinced Gold I was a glowering sociopath. In the lunchroom I was as social as anybody, doing impressions of the instructors and railing against the general ridiculousness of tower life. But apparently I'd been better than most at buttoning up quickly when an officer came into the room.

I assured Gold that everything was fine and that I would try to show my "spirit" more openly. "Because if there's something wrong," Gold said hesitantly, "we can get you into counseling or something." The idea of counseling was so ludicrous to both of us that he let out an audible sigh of relief when I rejected his offer. It was clear that he was following the rules of some sort of "principles of supervision" class that management had made him take, and he was entirely uncomfortable in the role. He shook my hand and sent me out. From then on I made a point of shouting an enthusiastic, "Good morning, Captain Gold!" whenever we passed in the halls. Occasionally I'd even let him catch me telling a joke and then do the inevitable punishment push-ups that followed. It must have been what he was looking for, because he never mentioned my psyche to me again.

When I left Gold's office, I was jittery and nauseated with relief. I hadn't known how much I wanted to keep the job until I felt as if I might be losing it. I was thrilled also by Gold's stab at introspection. His awkward "pull yourself together and for God's sake let's not talk about our feelings" routine was the perfect antidote to a lifetime spent sharing and overanalyzing. Just black and white, no gray whatsoever. Water in, water out, dick to the fire and we'll all be just fine.

That day at lunch, everyone gathered around while I told the story of my run-in with Gold. Steve Krale, a giant bodybuilder who had already worked in a different department and could never be accused of being too serious, thought the story was hilarious. "I always knew you were unstable, you little freak. Everybody watch your backs when Prozac's around—he could snap any minute!" For weeks afterward he would cower in front of me in mock terror. "Please don't hurt me, Prozac, I swear I'm your friend." Prozac: The nickname stuck.

The fifty-foot ladder is a relic. Its primary remaining purpose is to terrify new kids by being heavy and dangerous to throw. By the time of my training, most of the truck companies in the city had long since retired these big sticks in favor of mechanized aerial ladders. But there was still a fifty-footer lurking behind one of the trailers at the drill yard. When Gold stormed out of roll call one morning snarling, "I'm tired of this little shit; let's throw the fifty," we all knew enough to be scared.

There are three kinds of ladder in the fire service: straight, extension, and aerial. A straight ladder is what you think of when you think of a ladder: round wooden rungs between wooden rails. Fire-department straight ladders are different only because of the sharp feet at the bottom end, ragged edges of metal hard enough to take a bite out of asphalt as they dig in for a solid grip. Straight ladders are the tools of choice for the sake of simplicity; they can be thrown quickly in narrow alleys or on uneven ground by one or two firefighters. It seemed odd to me to use wooden ladders around a fire, but nobody ever mentioned it, and I knew better than to ask.

Extension ladders have two sections, one nested inside the other. The fifty-footer is an extension ladder. After an extension ladder is thrown to a straight-up position, one firefighter grabs the rope halyard and hauls the second section up until it reaches the needed height. A pair of locks, called "dogs," click past every rung until they snap into place. If for some reason the dogs should fail to lock—say a firefighter loses his grip on the rope or there's a miscommunication between part-ners—the second section will come crashing down inside the first, nip-ping off whatever fingers or toes might be in the way.

Finally there is the aerial, the king of ladders. This is the ladder in "hook and ladder," the big stick on the big rig with two drivers, one in front, one in back. The aerial is fully mechanized. After blocking the wheels and screwing down the side-stabilizer jacks, throwing the aerial is as simple as pressing a button and moving a joystick. The driver is the one who throws the aerial, because the hardest part of the operation is deciding where to park in order to hit the target while missing light poles and power lines.

The fifty-footer is the longest of the hand-thrown extension lad-ders, so massive and unwieldy that it takes six people to hoist and one to direct. With staffing cuts over the years, there now are simply not enough people at a fire to be able to spare half a dozen men for a single throw. But no such restrictions exist in the academy, and the fifty de-mands the strength, nerve, and smarts that Gold was so fond of telling us we lacked.

"All I want to see here are asses and elbows," Gold said as we gath-ered around the enormous ladder lying on the ground. "You have to lis-ten to exactly what I say. This is not a joke here, and someone could get very hurt." None of us had any trouble believing him. By this point in the academy, everyone was downing Advil by the bottleful and not looking forward to any more pain. The fifty is so long that when fully extended it flexes dangerously and loses stability. To compensate, two long poles, called tormentors, hang down from the top of the first sec-tion in order to brace the middle of the ladder against the ground. The two firefighters in charge of the tormentors use them to help shove the ladder as it goes up and then set them into their stabilizing places after the ladder has been leaned into the building. We took our positions—two at the foot, two at the head, two on the tormentors.

"On my command. Ready at the foot?" Gold planted his feet, put his hands to his hips, and shouted.

"Ready, sir." Phan and Russell.

"Ready at the head?"

"Ready, sir." Antoine and Dwayne.

"Ready on the tormentors?"

"Ready, sir." Me only.

"Excuse me, Captain Gold? Am I holding this right? Can you please show me again?" said Mary Mills, my partner on the other tormentor. Everybody sagged. I didn't know exactly what I was doing either, but I would rather have had the ladder fall on my head than admit my cluelessness. We'd had our muscles tensed, our minds made up and ready to throw the thing, and now Mary had stalled the entire operation. We'd watched other squads fight the fifty, and nobody was looking forward to working with Mary on this tough throw. I could see where she was heading—she wanted her feelings massaged, she wanted to share. My vicarious embarrassment for her was so strong that I buried my face in my shoulder so I wouldn't have to watch.

"No, Mary. Put the point just between your finger and thumb so it won't spike through your hand when you push. Do it the way I told you," Gold said.

"It's just that it doesn't seem comfortable. I mean, that's not the way to hold something if you need strength. It's not working for me."

"You're not supposed to be comfortable, Mary. You're a fireman. It hurts sometimes." The word "fireman" hung in the air like an insult. Gold ignored it; he was so far beyond angry that he was already cool again. It had been a long day. "Let's try it one more time. Hold it like I told you to."

We readied again. Mary shook her head and muttered to me, "I just don't think this grip feels right for my body type." I ignored her.

"Asses and elbows, now. That's all I want to see, asses and elbows. All ready . . . now, *push.*" Gold gave the command. Three hundred pounds of ladder started to move. I could see the strain in their bodies as Antoine and Dwayne walked their hands up the rails of the ladder, taking the weight straight down on their shoulders and arms.

"Now, hold, everybody hold." Gold looked at the tip of the ladder, which now pointed skyward. A fine rain blew, and the wind was running fast, swirling differently at the tip of the ladder than at the base, where we were. The ladder leaned precariously, and we all struggled to keep from losing balance. "Fuck," said Antoine under his breath, speaking for all of us. With every gust of wind I had to decide whether to stay and fight the sway or bail out so I could be clear when the ladder collapsed on top of us. I wanted to run but didn't want to be the first to do so. I wondered if anyone else was thinking the same thing. The ladder leaned out, and we leaned back into it, overcorrected, and pushed the other way, felt the sharp feet of the ladder walking unsteadily across the asphalt as we strained.

"Tormentors . . . come around *now*."

I came around. Mary stayed put. "Sir? Captain Gold? The handout said that we're not supposed to come around until after we lay the ladder into the building. So which order should we follow? It's all very unclear."

"Bring it down! Bring the fucking ladder down right now!" Gold was screaming. I loved the idea of bringing the ladder down, but I knew we would just have to start the evolution over again from the beginning. We brought the ladder down in silence and shook out our aching shoulders.

Gold glowered at Mary for just a moment, then took his helmet off his head and kicked it. It skittered across the yard, spun for a second on its brim like a top, then came to rest upside down, with the face shield snapped off. All the other squads stopped whatever they were doing to stare at us.

Gold turned away from us, but from where I stood at the far end of the offside tormentor pole, I saw something I never expected. Gold smiled. Not a malicious smile or a sadistic wince. No, Gold was chok-ing back a laugh! The dropkick had been theater, and Gold was enjoy-ing his own act. It made sense, in an odd way. Since when had I been serious and silent, in possible need of psychiatric care? Only since I had begun my time in the tower. Maybe Gold was just playing a role as well. For the last few weeks, I'd noticed that the endless repetition of threats had been reducing my capacity for fear. I'd begun to look on Gold's ha-rangues as a welcome moment of rest from hose and ladders. Suddenly those fuck-you boots didn't seem so menacing.

But Gold regained himself quickly as he caught Burton's eye across the yard. He walked slowly to his helmet and stooped to pick it up. Mary stared at Gold, unchastened and pouting, one hand on her hip like a teenager who's just been grounded. "I'm feeling as though we're all having a hard time knowing what you want because you keep telling us different things," she said to him. "To tell you the truth, I don't think you've developed a very effective teaching style at all."

Gold's eyes bulged. "Your *feelings*, Mary? My *teaching* style?" What-ever it was I had seen in him was gone. Gold jammed his helmet back onto his head and stormed off, leaving us in the rain with a fifty that we hadn't yet managed to throw.

I wanted to feel sorry for Mary, I really did. But I was surprised to find that there was very little inside me but scorn. After all, hadn't I made it this far? She and I had been brought up with the same vocabulary about "process" and "individual style," but hadn't I been able to squash it? When the STAR drill kicked my ass that first week, I'd just changed my approach and come back at it, learning to take a slow pace and work steadily and with deliberation. I'd learned to watch Antoine and the other professionals, gotten over my old bad habits quickly.

Mary was right, of course: Burton was a bumbling goof, and Gold was a tyrannical ass. Every criticism Mary had raised over the last several months was right on target. We weren't learning anything from the eye-bulging tirades, and we weren't becoming better human beings by falling in line with ridiculous norms of obedience and discipline. I was chafing every bit as much as she was; when you've been brought up to believe that your opinion matters, it's hard to stomach becoming a nobody overnight.

But Mary never changed. She just felt that if she wasn't learning, it had to be Gold's fault. And that was probably true, but when it came to his authority, Gold didn't care about the truth. We weren't there to learn nuts and bolts. The firefighters' kids in the academy confirmed that once we were out on the job, we'd probably never see a fifty-footer again. We were there because every Oakland firefighter before us had been there, too, had taken the same abuse. We were there to be indoctrinated, to learn that at a fire somebody leads and everybody else has to follow.

To some extent it's true that Mary stuck out not only because of how she acted but because she was a woman. Everyone tried to pretend that they never even noticed gender, but with three women and twenty-one men, it was hard to ignore the disparity. But of the three women, it was Mary who was clearly the strongest physically—she was taller and in better shape than many of the men. And the other two women were doing great. LeAnn Calderon was an absolute fireball, the kind of person who was so intuitively competent that she could shout out tips and reminders to her partner without missing a step in her own portion of the job. And Deb Kinnear was quiet but solid, smart enough not to take anything personally, always keeping her eye on the distant goal of graduation. It wasn't Mary's woman-ness that was her problem; it was her failure to recognize that the academy was larger than she was, that becoming a firefighter was more than just taking a job and showing up for work. Unlike me, she had no inner chameleon and could never let herself just blend into the background.

I felt guilty, but I secretly approved of the way Gold yelled at her, the way he turned her pouty whines into curses he threw back in her face. Sometimes I felt like I should befriend her, help her through the academy, but it was so much easier to laugh with the other guys and tell stories about Mary at lunch. Though when she'd walk up and say, "What's going on, guys?" in a plaintive voice, I'd feel terrible. Years of circle time and cooperative Berkeley education were rapidly dying within me; all I cared about was my own survival. When in Rome . . . shut the hell up.

"Mary's not getting it," Antoine said to me as we were loading hose one late afternoon. I *was* getting it, though, and it was exhilarating. Despite no warm feelings for her, Antoine understood that we were being judged as a team, and her failure would make us all look bad. He'd convinced a few of us to stay after hours to give her some extra coaching. If anybody other than Antoine had asked, I don't know if I would have spared what little free time I had. There was only one week left in the tower, and our chances of graduating hinged on those last few days. We'd do the STAR drill one final time, be graded on our skills with hose and ladders, perform mock medical skills at a mass-casualty incident, and be subjected to a written test of all the information that we'd spent so many hours cramming into our heads.

At home at night, I'd enlist Shona's aid in practicing the complicated footwork involved in a three-person extension-ladder throw. To passersby glancing up into the lighted picture window of her apartment, we must have looked like stumbling tai chi devotees, determined to do the same step over and over again until we reached transcendence. I had enough to worry about without Mary and her shortcomings. But Antoine was right. Grudgingly I gave Mary a few hours of my time.

On the morning of the final manipulative exam, Mary was a wreck.

Her eyes were red, as if she'd been crying or smoking pot or both. I turned away when I saw her in the classroom, but she was oblivious to body language and made straight for me.

"I've been practicing all night," she said. "I think I've really got it. Thanks for all your help. You and Antoine both."

"No problem."

"How are you feeling? I mean, are you ready? Are you excited?"

"Sure, Mary, you bet," I said, edging farther away.

"I think Gold and I came to an understanding about how I need to be taught, about what the best way to work with my learning style is." It hadn't seemed to me that Gold was interested in understanding anything having to do with her "style." The last time I'd seen the two of them together, he'd been screaming and she'd been failing to hold back sobs.

"You just have to do your best," I said, offering bland encouragement and hoping she'd go away. I could hear Gold's boots coming up the wooden walkway outside the classroom, and I did not want to be seen with her, as if her manic, panicked excitement would taint me by association.

"That's a really good point, Zac. It's just like in sports, where you can only prepare yourself so much and you can't worry about the other player and—"

The door to the classroom flew open with a bang, smacking against the back wall. Gold stood in the doorway, hands on his hips.

"Sullivan, Phan—ladders! Unger, Earle—hose! Get your gear and get out there. This is the final exam, so don't fuck it up. Everybody else stays here and stays quiet."

Russell Earle and I ran outside and sat in the fire engine that was used for testing. We strapped ourselves into our air bottles, made plans for who would hook the hydrant and who would take the nozzle. It had all come down to one final evolution; whatever mystery fire scenario Gold threw our way would determine whether we'd be firefighters or just unemployed wannabes who'd wasted the last four months of our lives at hard labor. There were no second chances in the drill tower; if you failed, you were told to leave your turnout coat in the trailer and leave the yard immediately.

"Unger." The door to the fire engine opened, and Burton looked inside. "You two take off your shit and go inside. Hustle." Without offering any explanation, he slammed the door and walked off. We struggled out of our stuff and streaked across the courtyard at a hasty non-run that Gold always characterized as "walking with purpose."

My classmates were already assembled, and I took the last seat available in the front row. Captain Gold stood silently at the lectern in front of the classroom. He watched us for a long time before he began to speak.

"This is the final day of manipulative testing," he said, stating the obvious as a preamble to saying something that he found distasteful. "I haven't been pleased with any of you, but there's nothing I can do about that now. I brought you here to say that we fired two of your classmates today. They're finished. We don't need any dead weight around here. That's all. Get out there and finish up your tests."

He strode up the aisle of the classroom and let the door slam behind him as he left.

Phan Nguyen, the self-proclaimed last-ranked Number 24 of our class, broke the funereal silence. "I guess I have to be Number 22 now." With that, the room started breathing again. We all patted ourselves down as if to make sure we were still whole after a car wreck and then started shaking hands all around. There was a gleeful mass awareness that since we'd been sitting there to hear the speech, we couldn't be the ones who'd been fired. I could see everyone doing the same thing— looking around at the people in the room and ticking off a mental checklist of who was there and who was not.

The firings, while frightening, were not altogether shocking. John Hibbard—a slow, heavyset but unfailingly nice man who would have probably made a fine firefighter but couldn't manage to pass any of the written tests—was gone. He'd be missed.

And Mary, of course, who wouldn't. They'd been building a paper trail on her for weeks, documenting her outbursts, her continual failings to conform and perform. Antoine, who'd worked the hardest and been the most generous in helping her out, said to me, "I'm glad she's gone. There's no place for her in the fire service—you'd never see her anywhere close to a fire." Mary wasn't a bad person; with all her ideas about personal development and body position, she probably would have made a hell of a fitness trainer, which is what I assume she went back to doing. But Antoine was right: At this stage of the game, there was no room for individuality.

I'd shed my flaky hippie skin and ducked into a heavy turnout coat unscathed. I was surprised to find myself loving Gold's paramilitary rigor. Everything was so clear, so black and white, so fire and water. With Gold I always knew exactly where I stood; there were no gray areas, no problems to be worked through, no feelings to be endlessly processed. Dick to the fire. No questions asked or accepted. Mary was a victim of "Kumbaya," whereas I had remade myself, out of both fear and necessity, into someone stronger, less flexible. Someone more like a fireman.

When Gold barked at me to get the hell back out to the rig and finish my testing, the only thing I could think was that his screams felt like freedom. The incessant threats of dismissal had apparently been real after all, and yet somehow I felt untouchable. Over sixteen weeks I'd become sure and capable; the ladders weren't as heavy as they'd once been, and all of the myriad couplings, fittings, and nozzles seemed to jump into my hands in exactly the right order. I was running the STAR drill with ease now, watching my pace, rationing the precious air in the bottle. I felt comfortable inside my mask, able to fight off my early panic and replace it with focus.

I finished the tests in a blur, spinning couplings and dragging hose just as I'd rehearsed for months. Russell and I performed "two leads into a standpipe," the standard procedure for high-rise fires that I'd seen demonstrated at the orientation so long ago. Later, Gold read out our names, not alphabetically as usual but in the order of our ranking in the class, which determined our seniority over one another. Antoine was first, and he nodded quietly at the congratulations, as if he'd known all along that he deserved the honor. Phan narrowly avoided being last, and he stamped his feet in mock disbelief. As for me, I ended up precisely in the middle—just the sort of faceless mediocrity I'd been aiming for ever since Shona and I stood at the top of the tower on that first day of orientation. I'd never been in the middle of anything before, not test scores or college admissions, but on that day it felt simply perfect.

· 4 ·

Danger Boy

Until I was nearly done with high school, my literary diet consisted exclusively of what my mom referred to as "Ritz Cracker books." Every time I'd bring a new one home from the library, she'd look at the cover and then pretend she was reading a summary from the dust jacket: "Shipwrecked in the Arctic Ocean in winter, three adventurers cling to a Styrofoam packing peanut for survival, subsisting on nothing more than a single box of Ritz Crackers and their own earwax. This is the remarkable true story of their ordeal."

These were the only things that interested me as a kid: shipwrecks, polar conquests, Himalayan climbing tragedies, and deep-sea-diving near misses. By the time I was nine, I had read everything Jacques Cousteau had ever written. He was my hero, not because he was a great scientist but because he swam with sharks. I had a picture on my bedroom wall of Cousteau with a knife strapped to each leg. I considered Sir Edmund Hillary a personal friend, and I could quote statistics about survival rates of members of the Donner party. I was proud of the fact that I had memorized the names and exact heights of all fourteen of the world's eight-thousand-meter peaks; in conversation I tried to slip in casual references to the proper place to locate an advanced base camp when climbing K2 from the southern approach. I had weighed the ethics of starvation-inspired cannibalism and made my decisions ahead of time in case a crisis ever arose. I never went anywhere without a book of matches in my pocket, just in case I had to make an emergency bivouac. I was, in short, an insufferable disaster-geek bookworm.

Eventually, of course, the books weren't enough. One glorious afternoon I won five hundred dollars for being the tenth caller and knowing the words to "Coward of the County," Kenny Rogers's unsettling ode to gang rape, and I spent it all on scuba-diving gear. I forced my parents to drive to a beach three hours south of Oakland and sit in the sand watching as I got certified as a pint-size search-and-recovery diver. Once, after I'd dragged myself ashore through a pounding surf, a woman walking her dog cornered me and pressed her finger to my wet-suited chest. "Is that your mother over there?" she demanded, looking past me to where my mom was sitting at the far end of the beach, quietly staring out to sea. I told her it was. "Well, you should just know she was crying when you went underwater. I hope you're happy with yourself." I couldn't help it: I *was* happy with myself—exhilarated, actually. And when my mom saw me standing on firm ground, she ran over to me, clear-eyed and happy, too, full of questions, wanting to know if I'd loved being underwater as much as I had always dreamed that I would.

When I was a thirteen-year-old, it was rock climbing and snow camping. It was, "Dad, I think I need some crampons and an ice ax" and "Why *can't* I hitchhike to Yosemite?" One day when I was fifteen, my parents finally gave in and dropped me off in a motel parking lot in Grants Pass, Oregon, where a bunch of poorly washed river rats gathered me into a van and spent the summer teaching me how to be a white-water guide. Looking out the back window, I could see my mom leaning against the car, biting her knuckle. My dad stood next to her, hugging her around the shoulders; from their expressions you would have thought they were watching their only child being marched off to the electric chair. But weeks later, when I arrived back home bug bitten, wind chapped, blistered, and ecstatic, the only thing my mom wanted to know was when could she come on a trip with me.

Maybe there was something missing in my sheltered world that made me curious about the edges of safety. But it wasn't necessarily the danger I craved so much as it was the thought of the rescue; even as a very young child, I would rig up complicated webs of string for my plastic action figures to rappel from. Since war toys were out of the question in a family that went to antinuclear demonstrations for Sunday outings, I didn't shoot my toy men; I just buried them in rubble so that I could dig them to safety. As I grew into my teens the only thing that changed was that I began to pluck my rescue fantasies from the headlines of the day. When Baby Jessica fell into the well, I wanted to be the guy dangling on a rope to pull her out. When the marine barracks were bombed in Beirut, I wanted to be the one pawing through the debris, the one to find the last survivor trapped but alive in some tiny pocket of air.

If my life's intent had been keeping my mother awake at night, fire fighting should have been an obvious choice. Yet on the day when, home after college, I saw a fire-department job announcement postered to the back of a bus stop, I dismissed it without a serious thought. Since college I'd been working a series of odd and isolating jobs: laying blasting wire on a Nevada mine site, watching elephant seals mate on an island in the Pacific, searching for peregrine falcons on cliff ledges in a national park. My vague plan was to apply to graduate school, get a master's in forestry, and eventually start a career with the U.S. Park Service or the Department of Fish and Game.

"But you really don't like plants that much, do you?" my mom asked when I was at home between jobs for a few days of hot showers and fresh food.

"No," I answered. We were sitting at the table, and I was watching her make up the massive grocery list she always prepared whenever I came home for a weekend. I've always liked the way she does that; she makes orderly headings on a legal pad for each day of the week, and her scissors fly over the coupon books. It's more of a battle plan than a shopping list.

"And you hate bird watching," she said.

"No, I *loathe* bird watching. They all look the same to me."

"And you never liked the fish you saw when you were diving. I think you only ever liked the fact that you might die while you were looking at them."

"What are you getting at?" I asked.

"Just that since you're someone who seems to hate *biology*, you might want to think about whether becoming a *biologist* is the right choice for you." She went on clipping coupons as if she were making small talk rather than giving me a life-counseling session. And she was right, too. I'd always barely endured natural science only because it gave me a way to be outside hiking in the woods.

"I know, but I want to be able to really *work*. I mean, put my hands to something instead of just sitting at a desk. I don't want to go to grad school, but I can't see any other way of getting hired on full time with the park service." The thought of being *merely* an outdoorsman didn't square with my own elitist desires to rack up the diplomas. My boss at the national park in Utah where I'd interned had been able to break the monotony of bureaucratic work with the occasional excitement of a backcountry rescue and the promise of fighting wildfires every summer. A job like his was the most fantastic thing my limited imagination would allow—parlaying an advanced degree into ten days a year of pure excitement.

My mom clipped her coupons faster, the scissors snapping in a dangerous, manic rush. I could see that she was bursting with advice, that it was physically painful for her to restrain herself from telling me everything she was thinking. Instead she laughed, as if a crazy idea had just occurred to her. "Did you see the announcement for the fire department up on College Avenue? Now, *that* might be fun."

"I guess. Whatever." It always annoys me when my mom is right. "What do you want me to do? Apply for a job I saw on a *bus stop*?"

She set down the scissors and gave me one of her raised-eyebrow, why-are-you-joking looks. "Well, why not?" she said, exhaling audibly as she failed to contain herself. "You'd get to do what you like every day."

"Yeah, I suppose," I said. I'd meant my comment about the fire department as a sarcastic snap to end the conversation. I couldn't imagine being the first person in my family not to earn a master's degree. I wonder if she thought then that I might actually take her advice. I wonder if she knew that one day she'd lie awake nights worrying that somewhere on the other end of Oakland, her son might be sweating down some long hallway choked with flames.

"You should at least keep your options open, kiddo," she said, her code for the fact that, though she was going to be quiet now, she wasn't going to let the idea drop.

"We'll see," I said, not intending to see anything at all. Joining the fire department was too much of a stretch in my mind. I was going to be a park ranger, both rugged and intellectual, striding through the forests in my green pants and broad Smokey Bear hat. Being *just* a fireman was out of the question. My mom was onto one thing, though: I was happiest when I was doing the things that made her think I'd probably die soon.

And sure enough, had it not been for something my mom knew about me, something she saw in my restlessness, I would never have ended up groaning underneath a fifty-foot ladder or running laps around the drill tower under Captain Gold's watchful eye. I was out of town the day the fire department made the application available. There was no mailing away for it, no phoning it in. I figured that I was far too busy to come home from Utah and get an application. But my mom had other ideas.

She marched herself down to fire-department headquarters, stood in the long line of men (and a few women) easily young enough to be her children, and stuck out her hand when she reached the front desk. The people handing out papers must have blanched at the sight of my grinning, white-haired mother asking for her application. I can just see her standing in front of the tables, letting on nothing, daring them to look at her sideways. And in good Oakland all-are-accepted fashion, I'm sure they handed her the paperwork without comment and groaned to each other as soon as she was out of earshot.

At first, the prospect of being a firefighter was little more than a lark, a nice idea, a fun fiction with which to tweak my college friends as they went off to law school. *Have fun at the investment bank*, I'd say. *I'll be sliding down the pole.*

I do have a vivid childhood memory of a fire that burned a neighbor's house. What sticks out most from that night—standing in the street in my pajamas, well-used stuffed bear in hand—was not the fire itself but the gathered crowd and the air of agitation and community. Mrs. Siah put a blanket over my shoulders, and my dad disappeared down an alley with a garden hose, followed closely by Art Smith, our neighborhood Defender of the Peace, a tall, skinny man who worked with my father at the hospital and spent his spare time planting and replanting his perfect, tiny square of lawn. He was the man whom I was supposed to call "if anything ever goes wrong," the man who once appeared at our door with a claw hammer in his hand in the middle of a wild, multiyard chase that ended with his tackling an overmatched crook carrying a stolen TV. My dad was fighting fire with Art Smith! I felt like a hero just for being his son.

The idea of the fire department grew on me slowly. Initially it was a whim, just one application amid dozens of others to graduate schools and national parks. When my mom sent me the application in Utah, it was good for a laugh. But I submitted it—what did I have to lose? From that day it took almost two years until I stood in a firehouse for the first time since a grade-school visit. The hiring process seemed interminable and capricious. Physical-agility tests, written tests, interviews, background criminal checks, personality tests. It was like being vetted for a position in the cabinet. I wouldn't hear a word from the city for months at a time, and then all of a sudden I'd get a letter telling me that I had to appear for some new hurdle the next day or be eliminated from the process. And, strangely, with each new jump, it started to seem like less and less of a joke and more like something I really wanted. I got off a Colorado plateau one afternoon and drove without stopping through the desert night to make it back to Oakland in time for a just-announced interview with a battalion chief.

I may have wanted the job just because it seemed so hard to get. The written test was held in a convention center; with our number-two pencils at the ready, we took the test in shifts of several thousand. Everyone in the room seemed to know someone else. They'd all met in college-level fire-science courses or made friends while testing for other departments across the country. "I made it to the chief's interview in San Diego, but then I never heard anything back," I heard one guy say. "That's all right," said the thick-necked man to his left. "I couldn't even pass Denver's physical."

The other aspirants around me were filled with such a purposeful zeal that it was hard not to get caught up. I hid the fact that I didn't have a clue what they meant when they discussed whether they'd rather work on the engine or the truck or that they liked Oakland's "aggressive interior-attack" style of fighting fire. I could figure out the basics, though; they were the same as they had been when we visited the firehouse in grade school. I knew that firemen put out fires. I knew that it was dangerous and exhausting. I knew that they worked hard and that they always seemed to be laughing when I saw them in the store or washing the rig on the sidewalk. I knew that they saved people's lives, and once I started thinking of myself in that role, no matter how far off it still was, it seemed impossible to imagine doing anything else. How could I go back to wandering blithely through the woods when there were people trapped in burning buildings? How could I sit on a rock and watch some bird do its thing when there were parents who needed me to carry their babies down ladders before I wiped the soot out of my face and said, "Just doing my job, ma'am," then walked back into the smoke? The imaginary hero that I had been as a kid was getting antsy, ready for the real thing.

Whoever said that a man in uniform always looks good obviously hasn't spent much time looking at me.

With the written and manipulative testing completed, the last week of the academy was devoted to filling out paperwork, learning a few final details about proper firehouse behavior, and trading in our filthy work pants and T-shirts for the pleated dress slacks and dark button-down shirts that are the uniform of a professional firefighter.

At the uniform shop, I stepped into a changing room and tried on the outfit. It felt like a costume. I straightened my belt with the giant brass OFD buckle and stepped out into the store.

For a second everyone was silent, staring at me.

"Zac, you look terrible," Antoine finally said.

"Thanks, buddy. I try."

"No. You don't understand," Antoine said. "You look *really* terrible."

"You're just saying that because I'm Jewish, aren't you, Antoine? Can't we all just get along?" Antoine laughed and slapped my back. Because of the similarity of our last names—Unger and Ugade—the cadre had spent the last sixteen weeks confusing us for each other. We'd taken to telling anyone who would listen that we'd been separated at birth and that it was only because of different diets that I'd ended up white and curly while he'd turned out black and bald.

"Now I've seen everything: a Jewish fireman. Where's your beanie?"

"I left it in the locker room on the hook next to your dashiki, Antoine."

"Well, shit, all I know is that you better try something different, because right now you look like a fucking milkman." Everybody laughed, even Gold, breaking into the most genuine show of warmth I'd ever seen out of him. They were right; I looked ridiculous. I'd gotten the last hat on the shelf, and it was a few sizes too small. It perched on my head like a bowler on a bear.

I tried out a few poses in the mirror, shifting the hat to the side, standing at attention, gazing heroically off into the distance. None of them worked. I still looked like a slapstick caricature of a fireman. Even my name on the coat looked wrong: FIREFIGHTER UNGER. It didn't exactly roll heroically off the tongue like the names of other firefighters I'd met—Mickey Quinn, Fitz Killingsworth, John Ironside. But it didn't matter. These were my clothes now, and this was my job. Foolish looking or not, tomorrow I would graduate, and the next day—my first day—I'd wear these clothes to the firehouse to begin this life I'd spent the last four painful months preparing for. So what if I looked like a milkman? I wagged my finger at the mirror and intoned, in a 1950s public-service-announcement voice, "Remember, kids, never smoke crack in bed—you could start a fire." Everybody laughed, and I laughed along with them. It felt good, fresh, like a new start. The tower was behind me; maybe I could let Prozac die quietly and start to be myself again.

On the morning of our graduation, at the final roll call of the tower, we snapped up sharply without even a warning from Russell Earle. Gold walked up and down the aisles of recruits as usual, picking at a loose fiber here and there, muttering to himself under his breath. When he reached me, he stopped short, looked down at my feet, and then stared me in the eye.

"Are you *fucking* with me?" he asked. His tone of voice was the same as usual, but he had the smile that I remembered from his bit of helmet-punting theater. "All this time, and you only get it right *today?*"

For once my boots were gleaming. The night before, I'd taken the subway to the financial district in downtown San Francisco and let a professional shoe-shine guy perform his magic. But for the scuffs on the inside of the heels where my gangly, turned-out feet scrape together, the boots looked as good as new. Actually, they looked better than new. "You *must* be fucking with me. There's no way you'd work that hard on your boots unless you were fucking with me." Gold shook his head in mock disgust and took his place facing us.

"In sixteen weeks I haven't managed to teach you anything," he said. "You are not where I want you to be. Some of you are good, some of you are better than others, and some of you will never deserve to work on any fire truck that I'm riding on. But you're all going out into the companies, which means that I've given you everything I have to give." Gold stood uncharacteristically still, not pacing as usual. His voice was loud, and he spoke slowly, letting his words sink in.

"I haven't taught you anything because we've got nothing real here to burn. All of those hose lays, all of those ladder throws—they haven't got a thing to do with fighting fires. The only thing a ladder does is get you to a fire; it won't fight it for you. All of the hose lays we taught you only get you to the front door. They won't put out the fire by them-selves. What you learned here is nothing. You're in the car, and you've turned the key, but until you get your first fire, you won't have any idea what it feels like to drive." Gold stopped and looked at us, holding his gaze a moment to consider each person in turn.

"Some of you are going to catch good fires on your first shift. For some of you it might take months. Don't worry. They always come. But remember one thing: After today I can't help you anymore. It's up to each of you to keep learning. It's up to each of you to *want* to learn from every fire you go to. You need to seek out people who know what they're doing and demand that they teach you. You have to be a constant burr in their sides. Pick their brains, follow what they do, make them drill with you every day. You can't ever stop learning. And"—he stood straight as a ramrod—"you should never, ever presume to think that you know everything there is to know about fighting fire.

"Check your bottle every day, and always wear it when it's smoky. Stabilize the ladder every single time. Test every roof before you walk on it. And look out for each other. I don't care who you are—you can't do this job alone." Gold paced to the flagpole, then walked back to face us squarely.

"Somebody in your class is going to die. I don't take any pleasure saying that, but that is the truth. It might happen in twenty years, or it might happen tomorrow. Whether it's a fire or a car wreck or cancer, somebody in your class is going to die from this job. And for God's sake, don't let it be you." He let his words linger for a moment. "You're dismissed. Go do housework."

Gold had spent the last four months trying to convince us—and himself—that he was an ogre. He'd done his best to put on the wardrobe of a marine drill instructor, but I could see that he didn't truly have it in him. Back when he came in, I'm sure he did it for the same reason that I was doing it years later: He wanted to help people. When you've devoted your whole life to saving lives and helping strangers, it's almost impossible to play the hard-bitten, drill-sergeant role convincingly.

At graduation we marched to our seats in formation, trying to cut crisp military corners and snap off the regulation salutes that Captain Gold had taught us just minutes before. I could almost feel my mom and dad withering a bit in the audience as they watched their sweet hippie baby looking for all the world like a little soldier. The trajectory of my dad's life had been shaped by a desire to keep himself out of uniform as much as by any affirmative desire to stay in school. Even though he already had more degrees than a thermometer, he had continued to pursue a student deferment with revolutionary zeal long after the last American soldier had choppered his way out of Saigon. For her part, my mom had prohibited me from signing up for the Baskin-Robbins Free Ice Cream Birthday Club because she thought that it was probably a secret tracking program designed by the Selective Service System.

When my parents came onstage to pin my badge to my dress uniform, my mom put her hands on my shoulders and looked me up and down. The closest they'd ever come to a group of men with badges was the time a troop of National Guardsmen with batons and tear gas had run them out of People's Park in Berkeley.

"I never thought I'd see this," my mom said. "But you look—"

"Terrible. I know," I said. "Try not to worry. They let us wear T-shirts at the firehouse." My mom was right, though: The uniform would always look ridiculous on me.

But the badge—the badge looked good. Some of my classmates were pinned with the same badges that their fathers had worn before they'd retired out of the Oakland Fire Department. Others, like me, were given brand-new metal shields and high numbers—like mine, 542—that nobody in the department had ever worn.

The graduation itself was hurried and somewhat subdued. We'd all received our assignments, and some of us would be reporting for our first day of work the very next morning. We were done being recruits, and now we were probationary firefighters—"probies." The next year and a half would be spent rotating from firehouse to firehouse, drilling every day, and trying to learn the things that can't be taught at a training yard.

Senior Man Cuts the Bread

How much time you got in?"

Timo had his hands full, so he jutted his round face in my direction as he spoke, the words more an accusation than a question. The smell of adobo filled the room; Timo's Filipino specialty sat bubbling on the stove in a giant cast-iron cauldron that covered two burners. I could tell that the question meant I was doing something wrong, but I couldn't imagine what. In full new-kid frenzy, I had set the table, folded the clean towels, and done a prewash on the cooking debris. I'd slapped a heavy cutting board down on the counter, slid the French bread from its paper sleeve, and pulled the serrated knife from the rack inside the cabinet door.

"I said, how much time you got in?"

"Two weeks. About two weeks, Timo." He knew, of course—I'd been at Station Nine for most of it.

"And you're cutting the *bread?*" Keeping his feet firmly planted in front of the stove, he leaned back at the waist and hollered into the dayroom, where the crew had their feet up, watching a ball game on the TV. "Hey, guys! Get in here. The new kid's cutting the bread!"

Reggie shuffled into the kitchen in shorts and flip-flops, rubbing a hand over the stubble on his bald head. "Oh, boy, Timo! I can tell already we got us a lotta teachin' to do on this one." He jerked a thumb in my direction before plodding around to where I was standing.

While Nine had both a lieutenant and a captain, it was immediately clear that the shift belonged to Reggie, who ruled as something of a benevolent dictator. He was a large man—a giant, really—with thick, meaty hands and eyes that crouched in a perpetual squint. He carried a big round belly in front of him, which forced him to lean backward to balance the weight. I had trouble picturing him scrambling up a roof, yet Reggie was the unquestioned leader both in the firehouse and on the fireground. I couldn't tell how Reggie had assumed his mantle of power; maybe it was through physical size, pure seniority, or the possibility that nobody else really wanted to be the leader in the first place. It was hard to imagine him having won some sort of *Lord of the Flies* power struggle, though; his lazy eyes just didn't seem like they had that sort of fire.

"Didn't nobody never tell ya?" Reggie said, grabbing my wrist with short, sausage fingers and gently lifting the knife away from me with his other hand. "Senior man cuts the bread."

His statement seemed ridiculous, but Reggie's voice bore a tone of profound disappointment. Even as I thought better of it, I heard myself asking how using a bread knife could possibly be a seniority issue. Reggie squinted one eye shut and popped the other one up to look at me. He heaved a sigh and shook his head slowly.

"Because it's a rule. That's the way it always was. Senior man cuts the bread. It ain't as easy as you think—you can't cut it too heavy or too thin, can't cut it too early or else it'll stale up. And if you don't leave the ends in the bag, how you think Timo's gonna make croutons next shift?" Reggie sighted down the knife, inspecting it as if it were some vital and temperamental piece of machinery. "Really, at this stage of your career, you shouldn't even be *thinking* about bread."

I was crestfallen. How could I have been so stupid as to think that I could actually cut bread? Sixteen weeks in the tower had so thoroughly conditioned me to the likelihood of my own failure that it was hard to place myself at enough of a remove from my surroundings to recognize their absurdity. Unfortunately, if Reggie didn't think he was absurd, then I couldn't either; the sting of doing something wrong was entirely real.

"You're Peter today," Reggie said by way of changing the subject. I didn't even want to ask what he could possibly mean by that.

"You're Peter today," he repeated as I stared. "Peter's on vacation. You're here to fill his spot, so you sit in that chair there." He pointed toward a chair on the far side of the table. "You're Peter, easy as that." He shook his head at my ignorance, as if he'd known since birth that "being Peter" meant sitting at the far corner of the table. I sat where Reggie had directed me to, trying to figure out what it meant to be Peter.

I had hoped that with graduation from the tower, the worst of the nonsense would be over, but it wasn't looking good. Probation is designed to keep the new kid on edge. In addition to daily drills, monthly tests, and constant evaluations and reevaluations, the uncertainty of being in a new firehouse every week keeps the probie on his toes. Every firehouse is just different enough to be unknowable to the new kid. My classmates and I were perpetually on the phone with each other, like a coffee klatch of overmuscled Emily Posts sharing tips on firehouse etiquette.

New kids are expected to arrive at the firehouse between six-thirty and seven in the morning, a full hour before the shift officially starts. This time is always spent going over the rig, opening and closing compartments, memorizing the location of every tool, so that when some grizzled vet you've never seen stumbles out of bed wiping sleep from his eyes and croaks, "Hey, new kid, where's the Buckner fitting?" you can say, "Driver's side, second cabinet, top shelf," without missing a beat. Before coming to work, I always made a point to spend ten minutes driving around the neighborhood. There's nothing more embarrassing than having someone stump you by asking about a street that's right in back of the firehouse.

My crisis with the cutting board narrowly averted, I watched Reggie slice great slabs from the loaf, and when he told me to, I rang the dinner bell. The men drifted in from all corners of the firehouse, pulled along by the smell of Timo's cooking.

I quickly learned that Station Nine was a firehouse governed by rules; the men thrived on routine and loved to watch newcomers fall afoul of their arcane code. Both the salad bowls and the heavy steel pitchers of water for dinner had to be placed in the freezer immediately following lunch, or else the new kid would receive a vicious verbal beating regardless of whether it was his day to cook or not. At a meal it was not permissible to excuse oneself until the frozen layer at the neck of the pitcher had cracked and melted just so. If someone was standing in the way while you swept the floor, it wasn't enough simply to ask him to move. Instead the only acceptable phrase was, "One time, lemme through *one* time, pal!" All the old-timers (as well as new kids desperate to fit in) engaged in a constant banter, a highly stylized call-and-response that was as predictable as the changing of the shift each morning:

"Nice cooking, pal."

"Hey, I had a lot of help. Couldn't have done it without the men."

"You're the man."

"No, you're the man."

"No, you're the man. Nice cooking, pal."

Despite repeated urgings to "go get you some, kid," I waited until everyone had served himself before dishing out my own plate for lunch. The offer to eat first seemed like a trap—I could just imagine the shit I'd take if I put myself in line in front of the senior guys who'd "earned" the right to eat first. The plates themselves were huge, more like giant porcelain serving platters than anything I'd ever seen people actually eat off. Nonetheless, each one was piled high, with hunks of expertly cut bread overlapping the edges. Giant piles of chicken adobo spilled off the sides, and the guys heaped on mounds of lettuce covered so thickly with dressing that it looked more like soup than salad. A television mounted on the wall above the table was tuned to a nattering, reality-TV judge show. Everyone on the screen was screaming and crying at each other, shaking their fists. The volume on the set was turned down too low to make anything out, but it was just high enough to be annoying, to add another layer of chaos to the general disorder of a firehouse lunch.

Over the general din, a firefighter named Big Al was discoursing about the stopping power of various types of hollow-point bullets. He wore a filthy watch cap pulled low over his head, and as he spoke, he poured glass after glass of juice from the pitcher, drinking each one in great gulps as spots of grape punch dribbled off his chin. Everyone else must already have heard his shtick about his years in the military, because they all quietly struck up side conversations or feigned total absorption in the courtroom nondrama on the tube. Since he obviously had his sights on me, I kept my head down and did my best to concentrate on working my way through the massive lunch. When I glanced up for just a second, Al saw his moment and pounced. "Hey, new kid," he said. "Do you know how to kill a dog? I mean, do you *really* know how to kill a dog?"

"No sir." I couldn't tell if this was going to be a joke at my expense, an honest tutorial, or the amiable ranting of a genuine lunatic.

"Good, then. Now, listen up. You got to use the pick point of your ax. Don't even fuck with the flat blade part—don't even *fuck* with it, y'hear? Just find a spot between his eyes, go up an inch or so, point the pick at him, and send it home. Put your whole back into it, and send that motherfucking pick *home!*"

It wouldn't be the last time in my firehouse career that I'd find myself unsure whether to laugh or sit tight, silently dumbfounded. Big Al affected an amiable somnolence, but when he made a point, he showed bulging animal eyes and an openmouthed grin, both of which would disappear so quickly as to leave me wondering if he'd even smiled at all. There had been a rash of pit-bull maulings in Oakland, though, so I was inclined to hear him out.

"You see, killing a dog starts in the brain, before you even see the dog. You got to *want* to kill the dog, have to be okay with the killing of the dog before you even meet that damn dog that you gonna kill. That way, when he runs up on you, you just kill him cold. And then get ready to kill again—because you know something about those dogs? They all got friends, and you gonna have to kill them, too. Kill 'em. Just kill 'em, and don't you ever feel sorry you did."

He leaned back in his chair, drained a huge glass of juice in one gulp, and started laughing, a big, hearty belly laugh that would have seemed almost Santa-like but for the monologue that had preceded it. Big Al reached across the edge of the table to lay a hand on my forearm. His fingers stretched from near my wrist to just above the elbow. "Don't worry about me, new kid," he said. "I'm just a little guy trying to make it. I wake up most mornings feeling scared." He paused for a moment, but before anybody else could take up the slack in the conversation Al affected a professorial tone, complete with faux British accent, and said, "Now, where was I? Oh, yes, I prefer what they call a 'modified cop-killer'-type round for maximum penetration of body armor."

. . .

After Big Al trailed off and fixed himself on the TV in a reverie from which he would not emerge until well after all the dishes were done, the conversation quickly turned to everyone else's favorite topic: divorce. The men compared alimony payments and lawyers, debated over whether late-in-the-month overtime shifts could be hidden from spousal support. They wondered aloud why, if second wives are always better than first wives, do wives just keep getting better the more you have? I sat silent throughout dinner, uncomfortable, a little bored, but content for the moment to let ex-wives bear the brunt of the crew's abuse.

"Hey, new kid, you married?" Reggie shouted across the table during a lull in the conversation.

"No, not yet," I said.

"What's the matter—you gay?" Everybody burst out laughing. I'd already heard that joke three times before lunch at Nine, and I couldn't imagine they still thought it was funny. Just part of the routine.

"No, I'm not gay," I said. "Did you want a date?" I could see eyebrows arch at my comeback. It was a bit early for a new kid to get mouthy, but it was out of my mouth before I could catch myself.

"I'll bet you got a girlfriend, though," Reggie said. "Did you meet her before you got this job?"

"About a month before," I answered.

"Well, there you go," Reggie said, seeming satisfied with himself. "She knew you had a good job coming, so she sunk her hooks, and now she's reeling you in. Fish on!" he shouted to general laughter. "Whatever you do, don't let her marry you. At least you gotta sign a prenup so she can't run away with your pension." Shona was working as an attorney for a fancy downtown law firm, and I couldn't imagine that the prospect of a fireman's pension in thirty years was much of a romantic motivation for her.

Reggie continued. "I know how it is: You're loving her now, and it's all great when you come home in the morning. But pretty soon it's gonna be, 'Hey, honey, can't you stay at work an extra day? I wanna hang out with my girlfriends.' Don't you let her fool you. It all goes bad, way bad, after she gets that ring on her finger. Isn't that right, guys?" Everybody mumbled agreement around the table. I didn't say a word. I didn't want to take relationship advice from guys who thought that using a bread knife required a twenty-year apprenticeship.

I thought the ice on the pitcher would never melt. Finally it went with a crack, and they all pushed back their chairs. I took up a spot at the rinse sink next to Phil, a short man with round cheeks and nervous, shifty eyes. He had to stand on the tips of his boots in order to fish a fork out of the bottom of the deep sink. I was barely managing to keep up with Phil as he fired vaguely clean plates and bowls toward my rinse area. At one point he stopped chattering suddenly, brandished a large knife, silently held my eyes with his for a second, and then yelled "*Sharp!*" at the top of his lungs.

That seemed obvious. It was, after all, a knife.

"Thanks?" I said, unsure of the appropriate Station Nine response to this particular situation.

"You have to say 'sharp' when you pass a knife," Phil said, seeing how lost I was. "Otherwise someone might just reach out and grab it. This way nobody gets hurt. Sharp!" He handed me another knife. I rinsed it and placed it in the dish rack.

"Hey, Unger!" Reggie screamed at me. "Didn't Philly just get through telling you to yell 'sharp'? How come I didn't hear anything out of you?" He stared me down with thinly veiled contempt in his eyes, as if he were trying to decide if I was even worthy enough to receive his teachings.

"I was just putting it in the rack," I said. "And since Phil said 'sharp,' I thought I didn't have to."

"Every time, new kid, every single time. I could have gotten cut there."

I'd made it through my entire life without ever having been shanked by a dish rack, but I bit my lip instead of telling Reggie to just watch where he was putting his damn hands. A long day was getting longer. I couldn't imagine an entire career of this already stale routine, couldn't imagine myself at fifty being genuinely upset when some piss-ant new kid tried to steal my God-given right to cut a loaf of bread. For the moment it was easier just to give in. I decided that for as long as I was assigned there, I'd let Station Nine happen to me and try to learn as much as possible from these senior guys. I didn't want to get a reputation as the kid who wouldn't yell "sharp." I'd remade myself for the sixteen weeks I had to spend with Captain Gold, so it was only to be expected that I'd have to change myself again for however long I'd be stationed at Nine. Eventually I'd be running around the firehouse calling everyone "pally" and "partner," screaming "one time" and *"sharp!"* like a parrot with a severely limited imagination. Screaming just like the rest of them.

I wasn't destined to leave the dish line so easily, though. I'd gotten into the rhythm of yelling "sharp" for every butter knife that came my way, and I thought that maybe, just for once, I might not be the center of attention. Reggie had other ideas.

"So where you from, Curly?" he asked, making a desultory attempt at conversation while we dried the dishes.

"I'm from here. Oakland."

"Did you go to Skyline High?"

"No."

"Oakland Tech?"

"No."

"Well then, where the hell'd you go to high school, son?"

I'd been dreading this question. I'd even considered lying and telling people that I was home schooled, or in the Witness Protection Program or something—anything rather than the embarrassing truth. I concentrated on scraping a bit of food off the bottom of a pot with my fingernail and mumbled the name of the private school I had attended, a place that in its literature liked to tout itself as "an elite private prep school nestled in the Oakland Hills."

"Oh, *no!*" Reggie yelled, slapping his thick hands down on the counter to make sure that the entire crew was paying attention to my upcoming humiliation. "We got ourselves a cake eater! Listen up, boys! We got us a cake eater right here at the sink!" He turned to me. "Have you ever had your hands dirty, new kid? Is this your first time standing at a dish sink? Shit, I bet you don't even know which end of the mop you're supposed to hang on to."

I smiled and nodded while the rest of the crew threw rich-kid taunts my way. What other response could I possibly have? *Yeah, but . . . you should see the size of my tool chest?* It all stung, but I knew that their real judgment of me would be reserved until I proved—or failed to prove—myself on the fireground. This was just an easy way in, a vulnerability they'd probe in order to see whether I could take a hit or not. And more than that, I couldn't really blame them for their derision; I hadn't appreciated my prep school much myself. It wasn't worth telling Reggie and the guys any of that, though. Better just to work hard, shut the hell up, and let the barbs float past. A new kid should be seen and not heard. That was easy enough to understand.

Fortunately for me, the cast of characters at Station Nine was never static. With nine guys on duty at any one time, there were always a few who were off injured, on vacation, or simply trading a shift with another firefighter. It was rare to have the entire crew intact, and I looked forward to the fill-in guys who moved through the firehouse. One morning the crew was depleted because several members had to attend specialized training or go for their yearly physicals. In order to have enough manpower to staff the rig, Tracy Toomey, a veteran firefighter who had moved up to one of the slow hill stations, was detailed down to Nine to fill in for a few hours.

Tracy was a solidly built man with the rough hands of a laborer, thick fingers with blunted nails and a gray accumulation of grease in the creases of his palms that no amount of washing could remove. As I was beginning my housework just after roll call, he came up to me and took the broom from my hand. "That can wait," he said. "It's not what you're here for." He led me out to the apparatus floor and walked around the fire engine, throwing open every compartment, sliding out every drawer. For the next three hours, we went over the rig piece by piece. Every tool started a story, every length of hose recalled a fire from a distant past, called up tactics honed and perfected over years on the job. A Vietnam veteran, Tracy also ran a successful welding business in his free time, and he spoke with a gentle authority born of years of hard and dangerous work. With a welder and a cutting torch, he was as much an artist as a craftsman, often inlaying intricate metal designs into gates and window frames. He had a self-effacing confidence that I both admired and envied. While the rest of the crew was camped out in the kitchen drinking coffee and telling lies, he made the effort to share his wealth of knowledge with me, a total stranger starting out in the business that had been his life.

The first few weeks of any firefighter's career are an impatient wait to start living the dream, and mine were no different. This was back before I understood that it was also a job, that there would be moments of frustration and grief, back when it was a fantasy that had come true but hadn't stopped being a fantasy yet. There was no way to know how I would react until the first bell rang, the first black plume came into view as the fire engine screamed around a corner. I had four false alarms before my first working fire—unattended pots on the stove, a spark from an electrical socket, steam from a shower playing tricks on a smoke alarm. Each call was tantalizing, and afterward I would reluctantly replace my ax, shove my air bottle back into its rack, and continue the long wait.

After a morning with Tracy, I was more eager than ever to get my first fire. I mopped at the floors listlessly, pointedly ignoring the common old-timer advice to be patient, that there's always a big one waiting for you right around the corner.

When the speaker crackled just before noon, I knew I had my first fire. The woman's voice on the 911 call had an edge of hysteria, a high-pitched panic when she said "my babies" that sent a chill into my gut. I upended the mop bucket in my excitement and ran off leaving a big puddle of dirty water on the floor of the bunk room. Tracy met me at the rig with a huge grin on his face. "Sounds like a worker, kid. Let's go!"

There are two kinds of hose on a fire engine: supply and attack. While a fire engine can carry five hundred gallons of water, that amount will last for only two minutes of heavy firefighting. The supply hose links the engine to the fire hydrant and unlimited water. Once water reaches the engine from the hydrant, the engine increases the pressure and pumps water out into the attack hose, which is used to fight the fire. The engineer drives and runs the pump, the officer controls the scene, and two firefighters are responsible for the supply and attack hoses. One firefighter has to act as "leadoff," the man who connects the supply hose to the fire hydrant. The "nozzle man," by contrast, pulls attack line off the rig and heads straight for the fire. After the leadoff man charges the line, he comes running up to help the nozzle man, but he often misses the thrill of the initial entry. Nobody wants to be leadoff.

As we drove to the fire, riding backward in our jump seats, Tracy pounded on the metal engine shroud to get my attention, jutted his chin out toward me, and smiled. "I'll take the leadoff, kid." I'd be up front, on the nozzle. It was a huge gift. Veterans often cling to their nozzle privileges tenaciously, and as Tracy hopped off to hook the hydrant, my biggest fear was that I would let him down, prove myself un-worthy of his generosity.

Craning around to see through the front window as the rig ap-proached the fire, I realized immediately that no amount of training could prepare me for the reality of what I was about to do. In front of me was a tall, thin, two-and-a-half-story Victorian house, classically detailed with gingerbread under the eaves. The rear window on the ground floor was blowing fire, and smoke was pouring from the attic. All around me was a frenzy of activity. Two firemen with Halligan bars, medieval-looking claw hooks designed for forcible entry, were prying open the padlocked front door. Three others were throwing a long lad-der to the peak of the roof. The lights of the assembled rigs blinked to-gether in a blur, casting murky midday shadows of firemen at work onto the surrounding buildings. The scene quickly narrowed down as I con-centrated on my task. I dumped 150 feet of preconnected attack hose in a pile at the side of the engine, ran the nozzle up to near the front door, then returned to untangle the spaghetti so it would follow smoothly as we made our way into the building. I was entirely consumed by process, engaged only in steps following steps, jobs to be done. It wasn't until I finally had the line unkinked and a mask on my face with fresh air flowing and I saw Tracy kneeling at the ready behind me that I thought about what I was getting ready to do.

My world was immediately a mess of flame. Just over the lip of the doorway was a child's image of hell burning fiercely at the rear of the building, red and white spits of fire licking up from a stove and spread-ing outward to curtains and cabinets. In the academy we had been taught about "safe" fires, fires that could be reasonably fought, as well as those that required immediate evacuation and a transition to a defen-sive, exterior attack. Kneeling in the doorway, I realized that I had no way to tell the difference. As if he were reading my thoughts, I felt Tracy's hand on my shoulder, pushing me forward. The pressure of his hand on my back was the weight of years of experience, the knowledge that this was a fire we could beat.

I knew that a level above me there might be people hanging from a window, begging for rescue. Between me and them was only fire, feeding thoughtlessly on the wood that was their floor and my ceiling. With a corridor of flame ahead of me, the groan of the heavy rafters in my ears, and Tracy's steady presence behind me, I started down the hallway, launched on the first terrifying, exhilarating crawl of my career.

"Don't put water on smoke!" Tracy yelled in my ear. I kept the nozzle closed all the way down the hallway, though it took some mental effort not to start spraying water as soon as I felt the heat radiating onto the exposed skin at the back of my neck. Fighting fire is a "whites of the eyes" thing: If you shoot water too soon, the only effect will be to upset the thermal layering and cause the smoke at the ceiling to bank down to the floor, destroying the last shreds of visibility and potentially causing the hot gases at the ceiling to be forced dangerously back down to firefighter level.

"Now!" Tracy slapped my back when we hit the threshold of the kitchen. I opened the nozzle, and water arched out toward the stove, darkening the flames more quickly than I would have imagined. Two beats later the heat hit me, delayed like the crack of a bat until the ball is already in play. I pressed myself to the floor until it passed, or maybe until I got used to it.

Rolling up on a knee, Tracy pointed to a corner of the room that was still glowing red. Reggie had fired up a big ventilator fan in the front door, and the smoke was running swiftly out the back window of the kitchen. It was getting cool enough to stand up almost straight, and Tracy and I moved into the kitchen, dousing hot spots and kicking debris to the side.

When we went outside to take a breather, I was vibrating with adrenaline. The fire had started on the stove and rapidly spread to the rest of the kitchen. Old Victorians are notorious for "balloon construction." There are no fire blocks in the walls, and flames can quickly run from basement to attic through open spaces between the studs. The crew after us had charged upstairs to the attic and extinguished the fire there, while the guys from the truck had cut a hole in the roof to let the heat and smoke out. I was exhilarated—I'd done exactly what I was supposed to do at a monster fire, and it had gone out in textbook fashion.

"Don't worry," Tracy said, scratching underneath his neck. "You'll get a real fire soon." I'd thought what had just happened *was* a real fire. "Kitchen fires aren't shit. You'll know it when you get a good one." Tracy looked at me almost apologetically. "Not a bad way to ease into things."

He tossed me a bottle of water that he'd taken from the rig. "Wash off. You're bleeding."

"Where?"

He tapped his own forehead, and I reached up to mine to find a thin trickle of blood from a cut just above my eyebrow. Tracy lumbered to his feet and grabbed an ax to go back inside and start taking the place apart to look for hidden hot spots. He grinned when he saw me trying to wipe the blood out of my eyes with the back of a dirty glove.

"Nice job, new kid. You got your ass kicked."

After lunch I was sorry to see Tracy go. But the guys had returned from the Training Division, and it was time for him to head back up to the hills. I shook his hand and thanked him, then set my jaw for the rest of the day with the Station Nine regulars. At Nine it was always a relief when the alarm bell rang. The close quarters of the firehouse encouraged the guys to create strange rituals beyond just the cutting of bread, like the drying of knives—also a job for the senior man, since the single swipe with a dish towel requires less work than the onerous chore of drying a spoon. But outside the firehouse, out where it really mattered, these men had something to teach me that I actually wanted to learn.

The fire department performs the random jobs for which there are no experts. Each day brings odd new problems, situations that no one on the crew has ever encountered. That day, in the awkward period after lunch—housework done, old guys drifting off for naps, new kids trying to look busy—we got a call to an apartment building for a "health and welfare" check. A resident had heard shouting from the unit next door, and, unable to get inside, he had called the fire department. We could hear muffled yells for help coming through the wall, but in the absence of an apartment manager with a master key, there was nothing to do but to give the door a horse kick and let ourselves in. Inside, in the bathroom just behind the front door, was a shivering, naked, thoroughly embarrassed young man with his finger stuck in the sink where he had been trying to dislodge a hair ball. He had struggled with the finger on his own for hours before giving up in frustration and starting to holler. Everything slimy within arm's reach had gone down the sink in his efforts to extricate himself. The sink was dripping with a pleasant-smelling goo made up of shampoo, toothpaste, hair gel, and Pepto-Bismol. He had even managed to tie a length of dental floss around his second knuckle in order to "pull from a different direction." With his head resting on a roll of toilet paper and his swollen finger padded with his wife's tampons, he had put his toiletries to the ultimate test, and they had failed him miserably.

"It is way too cramped in here. I need *space*," Reggie said, his voice booming inside the tiled bathroom. In no time we had the plumbing disconnected and the sink ripped from the wall. We walked the guy and his porcelain appendage out into the hall, had him descend the stairs a few steps, and rested the sink on the banister. Surrounded by firemen in full costume, standing with a towel around his waist and his arm raised up to his overturned sink, it was clear that the man wished he had decided to starve to death rather than endure the spectacle he was now treating his neighbors to.

With no real plan on how to proceed, another firefighter and I went down to the truck and stripped off any potentially useful tool. The poor guy blanched when he saw the armload of power saws, pry bars, and sledgehammers we returned with. Each firefighter's suggestion revealed his own unique personality: Reggie (six foot five and equally wide) wanted to "*smash it!*" Nervous little Phil wanted to transport the guy *and* his sink to the hospital. And the lieutenant was mostly just interested in taking pictures. In the end, unsurprisingly, smashing won the day.

With the porcelain chipped away and the pipe cut short, we were left with a more manageable problem, although the patient—whom we had all but forgotten despite his proximity to the accident scene— looked as if he might vomit at any time. He had stuck his finger down the pipe and then through an aperture in its wall, leaving the finger bent and utterly jammed. One of the bystanders helpfully suggested an amputation at shoulder level, but the patient had long ago lost his sense of humor, even if the rest of us found it sort of funny. Using the metal-cutting blade of a reciprocating saw, we were able to make a se-ries of small cuts, work a metal wedge into the opening, and spread the pipe just wide enough for him to withdraw his hand. His ordeal over, he sat looking dazed against the hallway wall amid the wreckage of his bathroom, porcelain chips and metal shavings littering the floor. Water was dribbling onto his bathroom tile where we hadn't quite been able to turn the valve off completely.

"Well, we messed that place up good," Reggie said, loading a couple of crowbars onto his shoulder. "Now he knows he don't want no fire-fighters coming to his house. We'll never see *him* again."

Firefighters aren't movers or construction workers. We're not really licensed to do anything in particular, and we don't guarantee our work. It's mostly about the action; cleanup is for someone else.

After lunch we were joined by Alan Diaz, a new kid like me who'd been off at a drill for most of the morning. Alan was good-natured and big, like a giant, black Mr. Clean with cartoonish muscles, shaved head, and a gold earring. He'd been a probation officer prior to coming to Oakland, and his eyes and stance alone made me think that he'd probably been able to inspire just the perfect amount of respect and fear in the kids he'd worked with. So many of the recruits who came through the tower with me were eager-to-please, easy-to-scare children, but Alan Diaz was a man, and he carried himself like one. I was always happy having another probie, someone to dodge probationary bullets with, and in the unpredictable, often ludicrous environment of Station Nine, I was glad to have at least one person I could count on to behave rationally. It may have been because Alan seemed so unfazed by the crew that they'd taken to giving him a particularly hard time. When I heard the thump of ladders against the building, I knew he must be out-side in the parking lot running through some drills, and I went outside to help him.

Alan was standing next to a ladder, dressed in full turnouts despite the heat. Phil was beside him in a T-shirt, apoplectic and red veined, a desperate picture of little-man anger. He held a cigarette awkwardly in his mouth, like a kid from the chess club who'd taken up smoking to impress the marginally cooler kids over in the audiovisual room.

"If I ever see you carrying a chain saw without a strap again, I will personally write you up!" Phil stormed in circles around Alan and the ladder, throwing his arms over his head and screaming. "When I was a new kid, I *never* cut corners like that. You had better shape up, and quick!"

Alan stood watching Phil, rocking back on his heels, not letting his physicality be threatening in any way. In private Alan was big and bois-terous, all smiles and good cheer, but, like me, he'd learned the value of being quiet. He shot me a glance across the yard while Phil stamped his feet theatrically. Phil was so involved in his anger that he seemed to be upholding both sides of a conversation, shooting down various excuses that he imagined Alan to be making.

"You might be a big, strong son of a bitch, but that's not going to cut it around here. You've got to be smart, and you'd better goddamned pay attention. Is that clear? Is that perfectly fucking clear?" The swear words were out of place coming from Phil's chubby baby face.

"I got you, sir. It won't happen again," Alan said smoothly, betraying nothing. Phil kicked at the ground and stalked off. I walked over to where Alan was standing, and we both watched Phil disappear into the afternoon darkness of the apparatus bay.

"I'll tell you what," Alan muttered to himself. "If I wasn't on probation . . ."

Silently I helped him lift the ladder and place it back into its bed on the rear of the truck. Alan's fists were balling up and releasing, balling up and releasing.

He raised his voice a little and eyed me conspiratorially. "After I get off probation, he'd better not say *shit* to me ever again. Are you feelin' me?"

"I hear you, Alan."

"Talking to me like he was never a new kid . . . I'm a grown man—where does he get off with that rub-your-face-in-it bullshit?" Alan was fuming, but he had too much pride in himself to let his anger burst out. "What's with this whole place anyway?" he said, gesturing broadly at the firehouse.

I pulled a lever, and the ladders locked into place with a heavy sound of metal grating on metal. I told Alan about the bread incident earlier in the day. He snorted, then wiped his sweating head against the sleeve of his T-shirt.

"I was starting to wonder about you," he said. "When I came in after lunch and heard you yelling 'sharp!' like the rest of them, I thought you might have gone over to the dark side." He straightened to his full height and crossed his arms over his chest. "You didn't go over, did you?"

"I'm trying not to. Sometimes it's easier just to play the game than to make a target out of yourself."

"Don't be like them. They might know something about fire, but you already know how to be a man."

I could feel my face turning red; I was ashamed that I had fallen prey to their patterns, and that it had been so obvious to Alan.

"Being a firefighter doesn't mean you have to talk shit about your friends or play those stupid new-kid tricks. Don't let them get you; we're out of here in a week," Alan said.

I wished that I'd had the confidence and presence of mind not to need to be on the receiving end of Alan's good advice. I wanted to be stronger. Actually, I wanted not to *have* to be strong, to be able just to be myself. Shifts at Station Nine lasted forever.

"You're right, Alan. You'll never hear another 'one time' out of me."

"All right then, and if you ever hear me calling somebody 'new kid,' you'd better let me know."

Alan stuck out his hand for that fist-to-fist brand of Oakland handshake that I'm never sure how to respond to, whether I should start with my fist up or down or just tap his hand straight on, knuckle to knuckle. After a few awkward feints on my part, we managed a facsimile of a fist pound. I might not have had the moves right, but I liked the sentiment, the feeling of Alan and me being on a team together, guarding each other against any slips toward soul-sucking Station Nine conventionality.

"It's a deal," I said. "Nice going, pal. You're the man."

"No, Zac, you're the man."

"No, Alan, you're the man."

"You know what?" Alan said. "Fuck it. I *am* the man."

· 6 ·

New Kid

After a month at Station Nine, I had reached a plateau of comfortable unpleasantness. I had most of the odd social patterns committed to memory, knew when to utter which inanities, and was even learning about fire fighting, though not necessarily enjoying my instructors' routines. During the endless hours around the dinner table at Nine, I had found myself craving unscripted conversation and a looser style. When the word came—as Alan had promised it would—that I was to be transferred to Station One downtown, I was both excited to be leaving and worried that every place in the department might be more of the same. At least I knew who cut the bread at Nine.

At a quarter to seven in the morning, I pulled off Martin Luther King Boulevard and onto Sixteenth Street, by the back fence of Station One's parking lot. I didn't have a firehouse key yet and didn't want my first impression to involve waking the guys up by ringing the bell. It was still cold, so I pulled to the curb and left my motor running. I was wearing my dress shirt with the shiny badge, and for once my boots looked decent—I'd just bought a brand-new pair, and the store-bought shine was better than anything I was capable of with my own poor skills.

"Hey fiiiiiremannnnn . . ." I heard a woman's strange, strangled, high-pitched voice and a tapping on the passenger-side window. I didn't want to have anything to do with her, but I was, after all, a servant of the community, and I didn't feel like I could refuse. I reached over and rolled down the window. She leaned into the car in a classic hooker pose, elbows on the door, ass up in the air, wiggling side to side. She smelled like stale cigarettes and cheap cologne, and I had a vision of getting busted for soliciting prostitution and losing my job before I even got off probation.

"Can I help you?" I asked, wondering what the appropriate opening salvo in a conversation with a prostitute might be: *How's business? Are those boobs real? Little cold out for a thong?*

"You got such pretty curly hair," she said, reaching a skinny, track-marked arm through the window and ruffling her hand over my head. "You want to paaaarrty, baby?" There was nothing about her that suggested anything remotely having to do with a party. I wondered whether I was supposed to be her first trick of the day or the last one of a long night. She dropped her hand down to the steering wheel, just alongside my white-knuckled grip. She had almost completely inserted herself into my passenger seat, her breasts threatening to spill out of her silver-sequined halter top.

"No thank you," I said, remembering my manners. "Thank you, though. No thank you."

"You new? You don't got to be afraid of Dominique." She waggled her long finger at me, scolding, then offered me her callused hand as if she wanted me to kiss it. I tried instead to give it a hearty shake while shoving her back toward the curb and out of my truck. Dominique was definitely the wrong way to start my first day at Station One. Unfortunately, nothing I said seemed to have any deterrent value with her.

"You go on to work, now—I know how it is. But tomorrow morning when you're done, you come and see Dominique, now, you hear?" She pointed at a tenement next door to the firehouse. "Just ask. Everybody knows Dominique."

"Great. Thanks. I appreciate your offer, but I'm not really interested in— I have a girlfriend, actually."

"Honey . . . please. Please, please, please," she clucked. She must have construed my excuse as an invitation, because before I could form my sputtering protests into a complete word, she pulled up the lock, opened the passenger door, and started to fold her tall frame into the seat next to me. "Lots of boys got wives. That don't matter none to Dominique." It looked as though she was going to settle in. I was dying. I thought about starting the truck and driving off before she could get in, but for some reason—the uniform, the job, my manners—I was unaccountably stuck with the desire to be *polite*.

"Please don't do that, ma'am. Please don't get in."

"Don't you 'ma'am' me—I ain't your grandmother." She looked old enough to be *somebody's* grandmother. "We're friends. Call me Dominique."

"Okay, Dominique. That's fine, but please get out of my car." I did not feel like her friend.

Just as she smoothed her nearly nonexistent leather miniskirt over her thighs and began to close the door behind her, we were both startled by the roar of a badly tuned exhaust pipe. A shiny powder blue Bug convertible with firefighter license plates skidded around the corner, top down, loud music blaring.

"*Dude!* What are you thinking?" The driver pulled up next to me and stood in his seat surveying the scene, smiling and obviously enjoying my discomfort. Dominique slapped her hands together in frustration. The driver was laughing around the cigarette in the corner of his mouth. He was young and Hollywood good-looking, but with the round belly of a former athlete who'd let himself go for a few years. I was so relieved I could have hugged him.

"Man, you are *new*," he said to me. I hadn't thought it looked so obvious. Turning to Dominique, he scolded, "What's wrong with you? Can't you see he's fresh off the fucking cabbage truck? Leave the poor kid alone."

Dominique glared back at him. "How do you know what he wants? I'm just trying to make a living. Why you gotta go and mess up a good thing? He was *liking* me." She had climbed out of my truck reluctantly, slammed the door behind her, and tried to hit me with one last come-hither look.

"Don't forget what I said about tomorrow, now, y'hear?" she said.

"Yes ma'am. I mean, no . . . ma'am. Just no. Have a nice day."

Dominique hitched up her breasts with great effort and swung off across the street, high heels clicking on the pavement. She made a bee-line for a silver BMW idling on the other side of Sixteenth Street.

My savior dropped back into his seat and hit the remote for the gate. He waved his arm at me, and I followed him in.

"You're not from Oakland, are you?" he asked, jumping out of his car and throwing the cigarette on the ground so he could shake my hand. There was that question again.

"Yes, I am," I said. "North Oakland, Rockridge area."

"Whatever," he said. His words rolled all over each other as if each phrase were in a race to be the first one said. "Same thing. You're not from around *here*. Welcome to Station One. I'm Billy. Don't let Dominique or any of those other guys give you a hard time—just ignore those dirty little fuckers. Fuckers: that's funny, huh? I am *hi*-larious."

"What do you mean, 'those guys'?" I asked.

"Holy *shit*, you're new!" he said, pounding me on the back. "What did you think that was—a six-foot-two chick with stubble?" Being propositioned by hookers before breakfast was so far out of my frame of reference that the idea that this particular hooker was actually a man had never even crossed my mind. I wished I hadn't been quite so naïve, but I couldn't help but be flabbergasted by the whole sordid interaction.

"But she had those breasts?" I asked. "She was falling out of that shirt thing she was wearing. He was, I mean."

"Dude, don't ask. I don't know how they do it either, but trust me, that's no woman. None of the girls who work around here are women. Skank city. This whole street is he-she land. Nice place to work, huh? Watch this." He pulled a new cigarette from the pack and threw it at his face like a suave fifties movie star. It hit his lip and bounced off. He picked it up from the ground, tried again, missed, picked it up, tried again. Finally he caught it in his mouth, and he flicked a match with a flourish and a grin. "Damn, I'm good. Shit, come on in. You can throw your stuff over there by the rig. There's coffee inside. I have *got* to take a crap. So what's your name anyway, new kid?"

I followed Billy inside and dropped my gear where he showed me. As I'd done at Station Nine, I started going over the rigs. Station One was a huge place, with a truck, an engine, a heavy rescue unit, and a battalion chief's car. By the time the bell for roll call sounded, I had barely made it through the cabinets on one side of the truck.

In the watch room, Billy was holding court over the rest of the crew, unself-consciously telling a story about being brutally rejected by a girl he was chasing. I introduced myself all around and was surprised not to get any of the immediate condescension I'd received on my first day at Station Nine. Captain Powell, a slight, red-haired man, stood holding a clipboard.

"Weeeellll, let's get started here," he said, trying to break into Billy's monologue. Billy ignored him.

"So there I was drunk off my ass, horny as hell, and wearing my underpants on the outside of my jeans, and now she's pretending like she's never even *met* me."

"Can we get started here, Billy?" Powell rubbed at his temple with the eraser end of his pencil and gave a frozen little smile when he spoke, as if he were reluctant to assert his command.

"I tell you what, though," said Billy, "it's not the first time I've been hung out to dry, and I'm sure it won't be the last. I've got another date tomorrow night. . . ." Everybody in the room laughed except for Captain Powell, who shuffled his papers and did some ineffectual throat clearing. Billy turned toward him with a look of false innocence. "I'm sorry, Cap—did you want to say something?"

"Thanks, Billy. Weeeellll, let's get started. This is Zac," Captain Powell said, repeating my name and introducing me as if I were the new kid in kindergarten. "Let us know if you need anything, Zac. Tom's been here for a while, so he can probably answer any questions you have."

"Oh, please!" Billy yelled from the corner where he'd been bobbing and fidgeting while Powell talked. "You're telling him to talk to Tom? That old smelly gorilla can barely tie his shoes without my help." He turned to Tom, who was slouched in a half-broken office chair, balancing a cup of coffee on his belly. Billy said, "Pops, did you forget to shave your forehead this morning? Because your hairline is advancing, and it's really not attractive."

Tom sat motionless, one hand on his coffee cup, the other slipped under the waistband of his pants just behind his belt buckle. He worked his jaw and sucked at his teeth as if he were trying to get the taste of sleep out of his mouth. He didn't acknowledge Billy and seemed to be staring intently at a spot on the floor.

"Hello! Tom! Over here," Billy said, snapping his fingers to get Tom's attention. "Is he still breathing?" Billy faked alarm.

Finally Tom lifted his head and swung his eyes over to the side of the desk where Billy was bouncing from foot to foot like a jogger held up at a traffic light. When Tom finally spoke, it was with a low, slow somnambulant rumble. "Billy, Billy, Billy. It's too early in the morning for your bullshit. Can I at least have my coffee before we do *Romper Room?*"

"No problem, Pops, I can wait. Hey—be careful, your knuckles are dragging again." Billy jumped up and down, hooting and beating his chest like an ape.

"Whatever you do on this job," Tom turned to me and said, deadpan, "just don't be like *him*." He jerked his thumb over his shoulder in Billy's direction. "That's the best advice I can give you. Now you know everything I know."

Captain Powell watched the exchange with a barely disguised grimace. He'd yet to hand out the day's assignments, and he was trying, without much success, to hide his impatience. He looked like a substitute teacher who had decided that the best survival strategy was simply to let the little hooligans run amok and hope for the best.

Out on the apparatus floor, I shadowed Tom as he did his morning checks on the truck. Tom McFarlane was the senior man at Station One, though he held the title lightly, and other than his permanent seat at the dinner table, he didn't demand any special privileges for his years of service. For thirty years he had driven the truck over the same ten-block neighborhood and watched hundreds of new kids pass through his firehouse. When I met him, Tom was still as thickly muscled as he'd been the day he arrived in the department—he still put himself through a punishing Jack La Lanne–style workout every day, involving massive iron barbells and chest-expanding spring pullers. He had a hooked nose, jowls, and large bags under his eyes. Billy wasn't entirely wrong in comparing him to a gorilla. Along with his muscles, he had a heavy forehead, long arms, and hairy knuckles. Tom had affectionately earned the nicknames Old Silverback, Pops, and the Missing Link.

Tom's charm was that he was utterly without artifice, as eager a worker as people half his age. While others might let minor details slide over time, Tom's routine never changed, and he continued to take the small but crucial safety steps that young firefighters quickly decide are beneath them. He didn't look cool, but he always wore his gear the way the manufacturers recommended, without any of the cavalier flair that so many other firefighters adopt. His turnout coat always fastened to the top button, his helmet strap always buckled and tightened under his chin, Tom was never more alive than when he saw flames dancing out of an upper-floor window. Most of his original academy-mates had eventually headed for the quiet of a hill station, but Tom had remained at a busy downtown house. Despite being a quiet man, he loved the give-and-take banter of an active firehouse and couldn't imagine working at a slower hill station, where the biggest excitement of the day might be watching the deer graze on the blackberry bushes.

Tom's combination of experience and strength was ideal for the types of fires common to a major urban downtown district. Fires downtown are rare because of the sprinklers, fire doors, and modern construction materials in most high-rise buildings. But when fires *do* arrive they are monstrous and demand every shred of a firefighter's intelligence and effort. In contrast, the mostly residential eastern end of Oakland has no high-rises to speak of, just tidy two-bedroom bungalows that burn with surprising frequency. Firefighters on the east end get the constant practice of kicking in a door, taking a saw to the roof, putting water on flames. But small houses have small fires, and they go out quickly. Only the guy who is first in gets to play, and the later-arriving crews often get there just in time to pull a few ceilings or drain hose in the street. A downtown blaze, by contrast, has enough fire for everyone. The mid-rises and tenements are our best bets for a good chance at some fun. Not tall enough to be well heeled and fireproof, not small enough to burn themselves out, and just old enough to be tinderbox flammable. A working fire in a five-story can take hours to knock. Often there are enough floors blowing flame that firefighters arriving on the second and third alarms still get the chance to grab a charged line and squirt some water. Downtown it is possible to go months between good fires, to be lulled into the security of false alarms and odor-of-smoke calls. But when fires come downtown, they come big. Tom lived for the big ones.

For years Tom and Captain Powell had sat next to each other in the front seat as driver and officer. They'd been to enough fires together to reach the enviable state of knowing what the other wanted without conscious thought when everything was going to hell. They made a perfect match, because Powell liked to micromanage and Tom not only liked to be told what to do, he could do just about anything he was told.

At Station One, I was the designated "bird dog," the position reserved for the least skilled member of a truck crew. As bird dog I rode facing backward in the jump seat immediately behind Captain Powell. At a fire scene, the driver and the tillerman—who also gets to drive the rear wheels—throw ladders and head to the roof. The officer and his bird dog are responsible for forcible entry and interior search. The fifth man in the downtown truck company goes where he is needed, usually to the roof with the ventilation team. This is the best-case scenario, of course. And a fire, by definition, means that things are not at their best: a woman hanging from a window and begging for rescue, a pit bull behind a razor-wire fence right where the foot of the ladder should be spiked, steel roll-up doors and mounds of garbage blocking the exits. This is perhaps the hardest thing to learn, that you can never learn everything you need to know for every situation, that a blueprint for putting out a fire is an imaginary construct. Instead over time you learn general strategies, suggestions, the capacities of your tools and of your crew. As bird dog I was just trying to acquire the general assumptions that would serve me throughout my career.

Powell was patient and tried to be generous with his time, yet even on the slowest of days, he seemed always slightly harried and nervous, as if the pace of a good running firefight had gotten under his skin and left a lasting infection. Years of successful fire fighting had made him set in his ways, though he always spoke in a retiring, almost apologetic fashion. "You really should have brought the bolt cutter on that call," he'd say, shaking his head as I unloaded ax, Halligan, pike pole, and the heavy extinguisher can that was slung over my shoulder. On the next run, I'd add the bolt cutter to my armful, only to hear him say, "Weeeellll, it sure would have been nice to have had a breaker bar up there on that last one."

Captain Powell was never at risk of being picked out of a crowd as a fireman. A quiet, unassuming man with a wispy flash of red hair who always seemed to be laughing at some private joke, Powell weighed 140 pounds after a good meal. With his air bottle on. A country boy at heart, he often spent his vacations bedded down beside the goats at some distant county fair where his kids were competing for 4-H honors. He lived out in the country, on property with a water supply so temperamental and sporadic that he filled up jugs at the firehouse before heading home each morning. The guys teased him mercilessly for his rural sensibilities, but he would just lower his eyes and smile, rarely firing back, comfortable with his place in the firehouse.

When he arrived in Oakland and zipped up his work boots, Captain Powell was undeniably a creature of the city. After twenty-five years he seemed to have seen everything and retained every bit of it. In a downtown crammed full of high-rises, Powell knew the location of every standpipe outlet and every fire panel, the average occupancy of each floor, and the particular hazards to be found there. He would go to the map before each run, but never because he needed to; I got the sense that he did it just to teach us new kids good habits. At the firehouse he was like a harried shopkeeper, scurrying along the hallways trying to keep every detail in his head while simultaneously appearing calm and professional for the sake of those around him.

Station One gave me a whole new outlook on the fire department. Station Nine was fast fading into a bad memory, and I found myself enjoying my day rather than watching the clock. Billy and Tom would tease me, but it was because they enjoyed being teased themselves rather than because they wanted to see if they could break me. Everyone at Station One did his job diligently, but without any of the mind-numbing posturing and rigmarole that I had feared would pervade the entire department. Granted, I was still a rank new kid, and the regular guys were trying to sound out whether I was someone they might want to work with or not. But, unlike at Nine, here I immediately felt like the rookie on a strong team rather than a pest who had to be driven out of the firehouse.

That evening a run came in for a car fire. Many Oaklanders view cars as communal property. Anyone who parks a car overnight in the flatlands is giving an implied consent for it to be stolen or looted. Even in her relatively affluent neighborhood, my mom never forgets to leave the ashtray open and in plain view, lest anyone think there might be a bit of parking-meter change worth breaking a window for. The pattern is simple: Somebody steals a car, drives it to wherever he needs to be, and then torches it to destroy any evidence. It's Oakland's version of public transportation. Anybody who has ever had a car stolen in Oakland knows that the idea of a cop's taking the time to dust a stolen car for prints is absurd, but it's better safe than sorry for the thieves, I guess, and a match is certainly cheap insurance against getting caught.

Rolling out of the firehouse, I looked over toward the bushes on Sixteenth Street just outside the cone of light thrown by a streetlamp. In the darkness I could see Dominique coming up for air in the middle of a blow job. "Hi, fiiiiiremennnnn," she called out, waving happily, as if we were having the most routine exchange imaginable. I turned my head quickly and tried to disappear into my seat. Billy tooted the air horn and waved back.

Somebody flying by on the freeway had seen flames and called in a vague address over his or her cell phone. I was on the engine, and we circled the neighborhood, smelling smoke and searching for the fire.

The truck—which was also in the area after coming back to quarters from a false alarm—found the fire first and radioed the location to us. Trucks can't do much at car fires—they carry nothing more than a few handheld extinguishers. Since it was a warm night and it's always better to watch a fire, any fire, than it is to go back to the firehouse, the truck stayed around to see the show.

In every good action movie, a car fire always leads to an explosion, a roiling ball of flame that sends sturdy heroes and their buxom sidekicks flying through the air. So that night when we got to the scene of my first car fire—Honda Civic, windowless, rag in the gas tank—I squirted at it from a respectable distance, aiming the water upward for maximum reach, letting the stream splash down on the blazing hood at the end of its long arc.

"Go on up in there," Tom called, hanging out the driver's-side window of the truck and waving me forward with a hairy arm. I stood rooted to the ground thirty feet away from the car. I had enough experience with James Bond movies to know that at any minute the gas tank would explode and I'd be riddled with flaming shrapnel.

"You can't put the fire out from there. It's burning from the bottom." Tom pointed to the flames issuing from the underside of the front end. The air smelled like burned plastic. I was perfectly happy where I was, but I moved a few steps closer.

"More," Tom called, sounding bored. I sidled closer, tensing for the explosion, feeling the heat on my face. Billy came up behind me with his weird little walk, like he was constipated or had bad knees, swinging his arms more than it seemed he should need to. He had a cigarette in his mouth and was wearing just a T-shirt, no coat or helmet.

"You gotta get closer, dude. I mean, c'mon . . . It's only a car fire." He made to grab the nozzle out of my hand, but I turned away. Already the acrid smoke was filling my throat, starting a burning cough from somewhere at the bottom of my lungs. It's possible that courage is only the desire to avoid looking like a fool in front of people you want to impress. I stepped in closer, kneeling down a few feet away from the car, feeling the heat puckering the side of my face.

Billy kept up a steady commentary in my ear, his forearm leaning into my shoulder. "Spray up under the wheel wells—that's where it's burning. Just a little bit. We're not hooked up to a hydrant, so don't make us run out of water. Knock it back, then wait and see where it flares up."

I opened the hose on the back, and the flames spit out the front. I put some water on the trunk, and the engine block lit off again, flames curling from the thin space between the car body and the buckling hood. Back and forth, flame to smoke, then back to flame again as soon as I took the water off. I was finding a rhythm.

Billy began flirting with a female cop who'd just rolled up. He wasn't particularly interested in her, but he couldn't help himself, couldn't stop exercising his biological imperative to get close to anything that walked upright and had breasts. He was smoking, laughing, lying, clearly amused, and using my timidity as a way to start his conversation with the cop. Emboldened by his easy voice behind me, I moved in closer, sticking the nozzle through a blown-out rear window.

When I heard the blast, my first thought was, *how could I have been so stupid?* I'd let myself be goaded into going too close, and now I was about to be rolled over by the inevitable explosion. I hit the ground and covered my head with my hands, waiting for the flames to wash over me and thinking about my upcoming life as a burn victim.

The sound of Tom laughing brought me back, heavy chortles followed by long sighs. I peeked up from between my hands. The car was still burning, but less so, flames visible only from the rear end.

"Damn, you *are* dumb," Billy said. "It was just a fucking tire blowing. It's got three more to go, too, so don't feel like you have to bail out every single time. What did you think—the whole thing was going to explode?"

Tom was out of the truck now and working on the hood, trying to find the mangled cable that would release the catch and expose the engine for us. He gave up and jumped up on the hood of the car, swinging his ax into the thin metal until he'd cut a wide V, exposing most of the engine block so I could wet it down.

"It's a finesse move," he said, walking back toward the truck. "You gotta be tough when you're stupid." Tom was anything but stupid, though this remained his favorite line; he wore it like a badge. "Nice job, kid. You'll get the hang of it."

I picked around inside the car for a while, searching for any hidden smoldering pockets. I trickled water down inside the doors, pulled out the floor mats, and ripped up the canvas ceiling. Billy was working on getting the trunk open, a standard practice to check if there are any dead bodies folded inside. I have yet to meet a firefighter who has ever found a body in the trunk of a car, but it's one of those things firefighters always check for, eternal pessimists that we are. Finding a body in a trunk is something that happens only in made-for-TV movies, but so much of what we see strains credulity to begin with that if there ever *was* a body in a trunk, nobody would be much surprised.

We stepped back and watched the car, trying to figure out if the white puffs were smoke or just steam lifting off the water on the hot engine block. Billy went back to the rig and coiled up the redline—an easily reloadable, small-diameter hose used for small jobs—while I stood staring at the steaming hunk of useless metal in front of me. Behind me, Billy hit the air horn, and I jumped.

"Are you just going to stand there admiring your work all day, or can we get out of here?" Billy called to me from the driver's seat.

"Coming." I headed for the rig.

"That was *awesome*, Zac," Billy said. "You are my freakin' hero."

Toward eight o'clock on my first night at Station One, dinner was long over and the crew sat lounging around the table. I was uneasy in my seat, looking at the stacks of dirty dishes, the cream sauce congealing into a hard crust. Three roasting racks sat on the counter, dripping with strips of glistening fat. Pots and pans of every size were strewn about amid little piles of flour and papery husks of garlic. The detritus of a normal dinner.

"One-pot meal, huh, Billy?" Tom said, surveying the damage. Earlier he'd piled his plate until the food threatened to run up his arm.

"You got it. One pot left on the shelf," Billy answered. Billy was a great cook, one of the best in the firehouse, but he never ran the risk of creating the much-sought-after and probably impossible meal that would dirty only one large pot and a single spoon.

After seconds and thirds were inhaled, the call for "lumber down!" thundered from one end of the table and was followed by the back-and-forth aerial zigzag of the toothpick container's being thrown from person to person. As a new kid at Station Nine, I'd been conditioned to believe that lounging was anathema, and on my first night at Station One, I couldn't stop popping up to get started on the dish work. Billy kept sitting me down. "Relax," he said. "Even prisoners get an hour."

I was sitting at the far end of the table from Herb Whitman, the on-duty battalion chief for downtown. Oakland is divided into three geographic districts, or battalions. Battalion chiefs—BCs—occupy the gray area between seasoned veterans and petty bureaucrats. They work a twenty-four-hour schedule just like us, eat and sleep at the firehouse, and respond to fires and emergencies as the incident commander. At a fire the BC will stand in the street and direct the attack, watch the big picture, and assume ultimate responsibility for the safety of all the firefighters, engineers, lieutenants, and captains working on the blaze. All the chiefs say that being a BC isn't what it used to be. In the old days, chiefs had absolute power; they could mete out discipline and transfer malcontents as they saw fit. Their primary job was to be the brains of a firefight, the steady hand that orchestrated the symphony of hose and ladder, ax and saw. Nowadays a battalion chief is welded to his cell

phone, his fax machine, his rule book. BCs today look gaunt and haggard, not from breathing smoke but from late nights writing memos, filing paperwork, and massaging bruised firefighter egos. A good chief
protects his men, acting as a buffer between the "workers" on the street
and the "propeller heads" in the administration building. The best
chiefs earn their reputations on the fireground by knowing when to
trust the men beneath them and by being able to spot the dangers that
men in the heat of battle can't stop to notice.

Whitman, with his thin white hair and rugged build, had defied all
logic of what should happen to a firefighter when he gets old. No persistent cough, no gimp in his gait. He'd worked for twenty-five years as
a firefighter and an officer, then turned around, became a BC, and
worked another full career at the chief's rank. With fifty-three years of
service under his belt, he showed no signs of slowing down. For over
half a century, he'd come to work every third day, like the rest of us, and
had his sleep interrupted by the bells at two in the morning, at four in
the morning, at six. Whitman had seen men he'd trained work full ca-
reers, retire, take long pensions, and die while he continued to prowl
the streets of Oakland, a fire-department radio permanently attached
to his hip. The rest of us might work downtown for a while, but compared to Whitman, we were all weekend guests.

"Hey, Herb," Billy called across the giant rectangular table. "Hey,
Herrrrb! What's new? What's the word on the street? C'mon, give it up.
Hey, *hey!*" Whitman pointedly ignored Billy and concentrated on picking a bit of corn out of his teeth. The two of them had a short but memorable history together, I soon learned, which was largely characterized
by the chief's pretending to be unamused by Billy's childish antics. One
often-retold story was about the time Billy was threatened with arrest
after a routine, good-natured chat with one of the neighborhood hookers. The rig had been stopped at a red light, and the prostitute—an
overzealous undercover rookie cop, as it turned out—had tried to haul
Billy to jail and impound the fire engine for good measure. Only Chief

Whitman's intervention had been able to defuse the situation, and for years afterward he would grinningly remind Billy of how close he'd come to a solicitation charge.

Tonight Whitman was making a show of ignoring him, but Billy was undeterred, like a three-year-old on a sugar high, with an insatiable appetite for attention. Billy kept up a ceaseless patter, talking to himself, to Tom, to the television, filling every momentary gap in the conversation with chatter. "Talk to me, Chiefy, I'm right here. Don't you want to talk to me? I'll let you have half my cake. . . ." Whitman stared straight ahead, concentrating on his toothpick. "What's happening to us, Herb?" Billy asked. "You've changed. We never make love anymore."

With that, Whitman finally broke into an almost childish giggle, then composed himself and turned toward Billy.

"Don't you have any respect for your elders? You shouldn't be calling me 'Herb.' I was putting out fires when you were knee high to a grasshopper, m'boy."

"Why do you say things like that?" Billy asked. "We already know you're old without you having to use those stupid old-guy lines like the grasshopper thing."

Whitman cracked his knuckles and stared at Billy. He pushed his glasses up the bridge of his nose with an extended middle finger, the extent of profanity that Whitman would ever condescend to use.

"Tell me a story, Herb," Billy said, shifting tactics and bouncing in his seat. "Tell me a story about the old days. Did you used to have to feed the horses?"

"Don't laugh," Whitman said, folding his glasses into his shirt pocket with just the slightest hint of a tremble in his hands. "I wasn't so far removed from those days, you know. Come to think of it, here's a story for you. Did I ever tell you the one about the dog?"

"Oh, *God!*" Billy said, rolling his eyes. "Do we really have to hear that same story again? You have *got* to get some new material, Chiefy."

Whitman nodded his head, trying to suppress a laugh. He knew that the story bugged Billy, but he eyed me and decided to tell it yet again.

"I was walking in the hills up by the reservoir—I do that every morning when I get off work. That's why I still look so good." He slapped his flat stomach and stared at Billy, caught midbite through an immense slab of cake, a bit of icing dangling from his nose.

"So I see this man coming toward me walking his dog. He's got this dog on a harness, and I can see it just pulling him along, really straining against the leash. Well, somehow it gets away and starts running. Runs straight at me. You should have seen the look on the guy's face—he's just horrified, thinking his dog is about to tear up some old fart. The guy's yelling at me. I can hear him while I'm watching the dog make straight for me: 'Climb a tree!' he's screaming at me, 'Climb a tree!'"

Billy pushed his plate aside, spread his arms in front of him, and thumped his head against the table. Then he grabbed Tom's arm and pleaded with him. "Make him stop, Tom. Make him stop! You speak Old Guy—talk to him. I'm *begging* you!"

"Shut up and listen, Billy boy." Tom slapped Billy's hand away from his shoulder. "Maybe you could learn something."

"Thank you, Tom," the chief continued, while Billy pretended to stab himself in the neck with a fork. "But fortunately I had my stick with me, a big old knotty thing. That dog ran up on me so fast that I had to just let him have it. I whacked that damn thing right on the top of the head. It gave me a little whimper and just kind of rolled over there and lay on the ground." Whitman was laughing at the thought. "Anyhow, eventually the guy catches up with his dog, and he looks down at it and up at me and back down at the dog. And he says, 'Now, what did you have to go and do that for? Why didn't you go up a tree?' I just picked up my stick, looked at him, and said, 'Well, sir . . .'" At this point the rest of the firefighters at the table—even Billy—joined in and shouted, "*I don't do trees!*" like it was a team cheer. Everybody was having fun, and I loved being in the middle of it, enjoyed how good-natured it was, how different it was from Station Nine.

Whitman continued. "So this guy's still looking at me like I did something wrong, and he actually asks me if I'll help him carry his dead dog out! I told him I had a walk to finish and he should have thought of that before he let go the leash." Whitman kept chuckling to himself at the memory. He shifted his attention back to Billy. "So, Billy, I guess what I mean is, I don't want to have to bring in my stick to work next shift. Do you think you can show me a little respect now?"

"Got it, Herb," Billy said with mock seriousness before breaking into patter again. "Can I go walking with you sometime? Just you and me, you know, like pals. I think we'll really hit it off, talk about the old days and stuff like that."

Just then Whitman's pager went off. He had a bad habit of getting called right when it was time to do dishes. At a firehouse dinner to honor his fiftieth year of service, the crew gave him a dish towel signed by everyone in his battalion. He held it up in front of him and said, "What's this for? I don't think I've ever seen anything like it."

"Don't worry about the dishes, Chiefy," Billy called as Whitman walked out of the kitchen, fumbling with his pager. "We'll just stay here and do all the work. It's no problem, really."

"Thanks, m'boy," the chief said. "Got to get an early start to keep ahead of clowns like you, Billy."

"Age before beauty, Herb. Age before beauty."

Blood Pressure

I've never been the kind of guy who cranes his neck at a car wreck in hopes of seeing some splatter. I can't stand it when somebody flips to some surgery channel on TV. Going through the emergency medical classes in the drill tower, I'd had no doubt about my ability to master the skills when it came to plastic dummies and written exams. Even the days when we'd perform fake mass-casualty drills with gory ketchup stains and plastic fractures protruding through the pant legs of our victims, I managed to do just fine. Yet I couldn't help but wonder if, after successfully enduring all the training and misery of the drill tower, I might just faint dead away at the sight of my first real gushing head wound. "I can't wait until I get to do some CPR," said one of my classmates. "They say you can actually hear a sucking chest wound *sucking*. That's gonna be cool." I wasn't so sure.

The only operation I've ever watched was performed by my father, who is a psychiatrist, which might say something about my ensuing aversion to all things medical. When my dad was still in medical school, our dog Scobie came down with an infected ear, which my dad decided to fix himself.

I can't say if it was because my dad was cheap or because he was cocky that he decided to lance and suture Scobie's infected ear by himself rather than go to a vet. He managed to cadge some anesthetic from the hospital where he was interning. With the help of a friend, he threw Scobie up on the kitchen counter. All the neighbor kids gathered around as my dad wielded his scalpel. I let them push past me to get a better view. I wanted to vomit and couldn't imagine ever eating my Cheerios off that counter again.

The procedure was a disaster. On the bright side, maybe it helped my dad reaffirm his commitment to psychiatry over surgery. I was glad that he didn't come home wearing bloodstained scrubs every night, wanting to talk about what the inside of a kidney feels like. For his part, Scobie certainly would have been better off had my dad stuck to asking him about his feelings. For the rest of his life, Scobie had a bulbous, partially closed left ear that swung off his head like a dangling clubfoot.

My queasiness with blood didn't get better as I got older. Somebody once told me that I have "white-coat hypotension." I've been known to faint dead away at the doctor's office, hardly the hallmark of a hard-boiled, seen-it-all emergency worker. The first time it happened, I had strep throat. The doctor asked me to say "ahhhh," stuck a tongue depressor in my mouth, and tried (unsuccessfully) to catch me as I crumpled to the floor. A few years later, when I strained a rotator cuff from too much swimming, I passed out after the doctor put his hands on my shoulder and worked the muscle back and forth between his two thumbs. Over the years my fainting habit had grown so commonplace that I would tell any medical professionals I came into contact with that when they did their evaluations, it would probably be better to have me lie down first. Sure enough, they'd slip the blood-draw needle in my arm and I'd feel darkness coming on fast.

It seemed inevitable that I'd see things much more gruesome than Scobie's ear over the course of a long career in the fire department. They'd told us in the drill tower that almost three-quarters of our runs would involve medical work, and I knew that my crew would have trouble accepting a queasy, fainting new kid as one of their own. As with fire fighting, there is nothing short of actual experience to reveal who has an aptitude and who will fail spectacularly.

"Gunshot wound. Corner of Eighteenth and Chestnut." I was on the rig before the speakers finished crackling. *"Engine. Engine. Gunshot wound."* On the way to the first major trauma call of my life, I smoothed my latex gloves over my fingers and tried not to think about vomiting. "That's a hot corner," Billy said. "Should be a good one."

The victim was sitting in his car, hunched over the wheel. He looked like a giant, folded into a tiny blue Toyota. "Pull him back. Let's see what we've got," Captain Powell said to me. I hesitated. You don't just go into somebody else's car. You don't touch somebody you don't know. "Go on. Get in there," the captain repeated. "You'll be all right." I opened the door of the car and pressed on the man's chest to move him backward. He didn't budge, so I leaned my shoulder into his and shoved him back into the seat.

The first thing I saw when I stopped to look was that the man's face was entirely missing. Under his shaved head there was just a mass of blood and gristle, nothing distinguishable inside the mess as an eye or a nose. His head slumped forward onto his chest, and I pushed it back again, my hand against his warm, spongy forehead.

"Get him out onto the ground so we can work him up," Powell said.

"Yes sir." I reached under the man's arms and around his chest. I hugged him close to me and eased him onto the ground. The captain put a hand under the victim's knees, and we carried him around his car, out of the driveway, and onto the sidewalk, where we had more room to work. The body was warm under my hands. Blood pooled on the ground next to his head, redder and thicker than I had expected. I tilted his head back and put the resuscitator bag over his face where it seemed like his mouth should be. Another firefighter was down on his knees, cutting off the blood-soaked tuxedo shirt the patient was wearing so we could start doing CPR. I concentrated on my job, on trying to somehow get oxygen through the guy's collapsed face and into his dying lungs.

When the ambulance arrived, we hoisted the patient onto a gurney, jumped in, and screamed off for the hospital. The other firefighter and I switched positions, and the loose bone ends of broken ribs made an unnatural sound as they grated against each other under the weight of my chest compressions. Blood ran down the sides of the gurney, dripping onto the floor, sloshing against the interior walls of the ambulance, streaming up and down in waves as we flew around corners and stopped short at intersections. I tried to maintain my balance and leaned into my job, concentrating on keeping my arms straight and counting out compressions: one two three four five . . . pause . . . one two three four five.

At the back door of the hospital, I climbed up on the rails of the gurney and rode it into the emergency room like I'd seen on television, pounding out my rhythm on the victim's chest as we rolled. I was wet from the exertion, and tried to duck my head around so that my sweat wouldn't drip directly onto the patient.

"Twenty-seven-year-old male," I said, pumping furiously as I gave my report to the trauma surgeon. He stood to the side and scowled, his arms crossed over his chest. "Gunshots to the face and abdomen. Downtime approximately twenty minutes. No pulses, no respirations."

The surgeon was silent for a moment. Then he glanced at the patient, glanced at the clock on the wall, and said, "Time of death: 11:23. Thank you, gentlemen."

Riding back to the firehouse, I experienced a kind of guarded exhilaration. I'd lost a patient for the first time, but I'd been in control throughout. I felt triumphant looking at the blood caked on the soles of my boots. I hadn't hesitated for a minute, hadn't had time to think about getting sick. I was so focused on what I was doing, so intent on counting out the rhythms of CPR, on doing my best to save a life, that I never wavered. He wasn't anybody I knew, and my only reason for being there was to try to make things better. Queasiness had no place in that situation, and, much to my relief, I'd held it together.

At the same time, though, I wondered if maybe there was something missing in me. This was the first murdered person I'd ever seen in my life, and it hadn't engendered any soul-searching. At the firehouse I tore into lunch as if I'd just been polishing brass all morning instead of watching someone die. I'd heard that emergency workers get inured to scenes of carnage, but I'd thought that it would be a gradual desensitization, not something that would happen on my earliest calls. Maybe firefighters don't *get* toughened—maybe we're all cold inside to begin with.

Emboldened by my success with the dead man in the morning, I jumped when the bells rang again during lunch. A chorus of groans went up when we heard the address, and the others pushed back their chairs, crumpling their napkins and throwing them on the table.

A sweet-looking old man sat in the lobby of a public-housing high-rise. He was nattily attired in a clean tweed jacket and a well-shaped bowler hat with a tiny red feather tucked into the band. He was sitting in a wing chair, and a cane rested on his knees. His hands trembled. He was crying.

"Frankie Hudson," said Captain Powell under his breath. Then he tucked his clipboard under his arm and stepped back, away from the patient.

"What's wrong today, sir?" I asked.

"My chest hurts."

"When did your chest pain start?"

"As soon as I started having trouble breathing," he answered, coughing into a folded linen handkerchief that he'd pulled from his breast pocket.

"Which came first, the chest pain or the shortness of breath?"

"They both came on when my stomach started burning."

The captain broke in. "What's the matter, Frankie? No headache today? You must be feeling better." He turned to me. "We picked him up twice last shift, and probably twenty times in the last month. He has to call from the pay phone in the lobby because the cops said they'll write him up for reporting a false alarm next time they get a call from his number."

I'd been feeling Frankie's wrist for a pulse, and now he grasped my hand and held it. He was shaking his head side to side vigorously. Tears rolled out of his eyes when he looked up at me. "Won't you help me? Won't you please help me, sir?"

"I'll do what I can. The ambulance will be here soon."

The captain stepped in next to me. "There's no point. He's like this every day. Just step off and let the ambulance take care of this one." Frankie was sobbing now, burying his head in the plush wing back of the chair he sat in. It seemed hard to believe that anyone would call 911 for the fun of it, and Frankie didn't look like he was having any fun. I wanted to help him, but the captain was right—I couldn't see anything wrong with him that I could fix. Frankie took off his hat and covered his face with it.

"Are you sure?" I asked.

"Just let him go. He always cries like that." The captain lowered his voice and leaned toward me. "He needs to get to the hospital for his morphine fix. You'd be crying, too, if you were as hard up for a score as this guy."

I don't think most rookie firefighters have any idea how few of the average day's calls will be genuine emergencies. The Frankie Hudsons of the world vastly outnumber the gunshot victims, the heart-attack sufferers, and the imminent childbirths. And not every chronic abuser is as sympathetic as Frankie. As in every major city, alcohol and drugs are the principal plagues afflicting Oakland's streets. One liquor store near Station One made a practice of holding people's welfare checks for them; customers would spend the month drawing down their balances in the form of malt liquor and fortified wine. We'd always find our regular "patients" sprawled on the pavement within a few blocks of the store.

Every firehouse has its regulars. On the east end of town, one particular clique came to be known as the "Food King Five" for their habit of hanging out in front of a run-down old grocery store. Every day at least one of the five—and often all of them—would call for an ambulance and demand a ride to the hospital. I'm not sure if it was the clean scrubs they were given to wear in the hospital or just the change of scenery that made them want to go for a ride. It got so bad that the crew who worked out there took to driving by the Food King and asking who wanted a ride just to preempt the call that would doubtless come at some later, less convenient time. One captain even offered Jerome Williams—the King of the Food King Five—a free one-way plane ticket to anywhere in the world. Jerome turned it down. He couldn't imagine a life for himself anywhere other than where he'd always been. The problem with chronic abusers is that, like anyone, they have genuine emergencies. No matter how many times a person cries wolf, the fire department can never afford the luxury of ignoring the alarm. Just before Jerome Williams died of a heart attack, he walked across the street and passed into the next fire district over. His regular crew never got the . . . what would it be? closure? privilege? satisfaction? of working him up on his final call to the fire department.

Oakland has what is known as a "multitiered" emergency-response system. When a person calls 911 for a medical emergency, the fire department is dispatched along with a private, for-profit ambulance company. The theory is that since there are more firehouses than ambulances in the city, the fire department can get on scene quickly, provide the first few minutes of emergency care, and then transfer the patient to the ambulance for the ride to the hospital. Often the ambulance crew can handle the problem on their own, but if it is a true case of life or death, then one of the firefighters will jump into the ambulance to help with the patient during transport. We have only one rig, so whether it's a high-rise fire or a headache, we're taking the entire engine when we go.

Years ago traumas were handled by hearses. Funeral homes, eager for business, would dispatch their long black cars to accident scenes. If the potential corpse happened to be alive, then the driver would aim for the hospital, but at least a seed of brand loyalty would have been planted should the patient (or more likely the patient's next of kin) eventually need burial services. The fire department is a natural fit for emergency medicine: We can be at any address within minutes, and we're used to dealing with crises, able to think clearly under pressure. In the Oakland of old, all the medical calls were handled by a battalion chief and his assistant. Medical-emergency calls were infrequent, and the treatment available was primitive; the main tool was the inhalator, or Flynn valve, a pressurized tank of oxygen attached to a mask used to force air down the patient's windpipe. With the popularity of the TV show *Emergency!* in the mid-1970s, the number 911 became cemented in the minds of Americans. It didn't take long for the chiefs to get tired of playing Johnny and Roy as they became overrun with inhalator calls. Now medicine occupies 75 percent of our time, and the same is true for emergency-service providers in most big cities. Sometimes it feels as though the other 25 percent is spent listening to old-timers grumble about how they never joined up to be doctors.

When I started in the department, I was trained as an emergency medical technician, or EMT. However, like many other cities, Oakland has recently begun to offer its citizens a higher level of care in the form of paramedics. When the city offered me the chance to go through a paramedic-training program, I knew it was something I couldn't pass up. Despite my initial fears of having a weak stomach, I'd come to enjoy the medical calls. Being cooped up inside the firehouse all day can be a little maddening, and I liked the chance to get out and interact with people.

The difference between a paramedic and an EMT lies in how deep below the surface of the body each can go. EMTs provide mostly palliative care; they can press gauze over a squirting artery or splint a shattered leg. They can even deliver a baby and use an automatic cardiac defibrillator. The role of the EMT is to hold hands, speak reassuringly, and in rare instances stave off immediate death until a patient can be handed over to a higher level of care. And in most emergency systems, that higher level of care is a paramedic.

A paramedic, unlike an EMT, can break the surface of a patient's skin. Since the medic has several dozen medicines at his disposal—high-powered drugs like epinephrine and morphine—he needs to be able to start an IV, to pierce a vein in order to introduce pharmaceutical salvation into a patient's bloodstream. Additionally, while EMTs can use a heavy rubber bag to squeeze air into an unconscious patient's mouth, paramedics can thread a tube directly down the windpipe until it lodges above the bifurcation that divides right lung from left. While procedures like chest decompression and tracheotomy are once-in-a-career rare, their presence in the bag of tools lends the medic a cocky swagger and a down-the-nose glance toward the EMT.

In paramedic school we were required to learn proper IV technique by practicing on each other. I let it happen once. As soon as some fellow student with shaky hands started swabbing my arm with an alcohol prep pad, I could feel the first familiar cold beads of sweat gathering on my forehead. I may have gotten good at dealing with the blood of strangers, but when it comes to my own body, I'm still a useless wreck. I'll start an IV on anyone, but if somebody points a needle at me, I'll be unconscious before they get the tourniquet on.

. . .

—

Being a paramedic prevents my job from ever becoming routine. I don't have the luxury of throwing up my hands and waiting for the transport ambulance to arrive. Not only do patients and their families look to me for help, but the rest of the crew defers to my medical judgment, accurate or not. One day, shortly after earning my medic license, I was working out on the treadmill in the firehouse when the bell went off. I jumped into my boots and turnout pants, toweling off as I headed to the rig. We were dispatched to an address I recognized but couldn't quite place. Pulling up on scene, I remembered that we had been there the week before for a man with terminal lung cancer who was having severe shortness of breath. I'd been able to hear the fluid gurgling in his lungs without even using my stethoscope; he had been blue in the lips and unable to speak more than two words at a time. This time he was lying in a freshly made bed, pulseless and not breathing. He had died peacefully, surrounded by family members who called 911 because they realized all of a sudden that they had a dead body in the bedroom and no idea what to do with it. The family produced a Do Not Resuscitate order, which specified that the man did not want any lifesaving measures taken. Unfortunately, the DNR had never been signed, and since the patient had no indications of rigor mortis, we were obligated by our local protocols to work him up. I clipped the laryngoscope together slowly; I was almost embarrassed to be holding it. This was a good death as far as deaths go, and the family didn't need to see the things we were about to put this dead man through.

CPR is an absolutely brutal procedure, and nothing like they teach you with those nice little resuscitation dolls the Red Cross has. It's also quite strenuous work if you do it right. Especially with old people, chest compressions tend to break ribs. You place your hands on the sternum right where you're supposed to, lean in heavily, and feel the crack-crack-crack when you push down. After the first few minutes of compression, the chest is a loose, shifting mess of broken ribs and torn connective tissue. Add to that the fact that patients invariably vomit as you're trying to ventilate them, and the glory of lifesaving fades out pretty quickly.

The worst thing about CPR is that it is almost never effective, yet we persist in this ritual flogging of the dead. There's always a chance, however slight, of bringing somebody back. But since we do CPR only on dead people—folks who are pulseless and not breathing—the rate of success is understandably low. CPR works best with people who are young and healthy to begin with and suffer some sudden offense to the heart, like electrocution or drug overdose.

Being without circulation is like being underwater. Brain cells burn out every second. Unless CPR is started immediately, a patient in cardiac arrest has no chance of survival, no matter how skilled the paramedics are or how quickly we can get to the hospital. If civilians aren't going to do CPR before they pick up the phone, they might as well call the coroner instead of 911. In the best of circumstances, it might take one minute for a bystander to realize there's a problem, one minute for dispatch to take the call, one minute for the firefighters to get out the door, one minute for us to drive to the scene, and one more minute for us to assemble our intubation equipment and start breathing for the patient. That's five minutes of being underwater.

And in reality it often takes much longer. Maybe there's a language barrier between the caller and dispatch, maybe there's traffic on the way to the house, maybe the patient is on the tenth floor and the elevator is broken. I've heard that Seattle has the best citizen CPR program in the country, that the procedure is taught to every high-schooler and city employee so that the chances of someone's receiving bystander CPR are greatly increased. It's the best place in the world to have a heart attack and the worst place to fall asleep on a park bench—because before you know it, three strangers will come over and start blowing in your mouth and pounding on your chest.

The best bystander CPR in the world wouldn't have made any difference on this call, though. Neither the family nor my crew had any hope that we could save this patient, and in fact none of us had any real desire to try. But protocol is protocol, so we dragged him out of bed and into the kitchen, where we would have some space. The captain and the driver from my crew began CPR while the other firefighter—also a medic—went to work. I lay down on the floor at the patient's head, pried his mouth open with my laryngoscope, and stuffed an endotracheal tube down his windpipe. The tube is more effective than a bag-and-mask system, because it delivers pure oxygen directly to the lungs with no leakage. While I was securing the tube, my partner had managed to insert a large-bore IV into the left arm. I called for epinephrine, one milligram, the frontline drug in most codes. I followed the epi with atropine, and we circulated both through the bloodstream with chest compressions. Every minute or so, I stopped all the action and rechecked the heart monitor to make sure that the patient was still in flatline. With enough epinephrine you can give a hamburger a heartbeat, but this man was beyond any hope no matter how many drugs we fed into him. We repeated the drug sequence one final time, but the patient stayed dead. Finally we flipped off the monitor and found a blanket to cover him up with. Only a coroner can touch a dead body after resuscitation efforts have failed, so we had to leave everything in place, including all the grisly medical devices that flagged our failed intervention. And all because he never signed his Do Not Resuscitate form.

"I'm sorry, ma'am," I said to a woman who might have been his wife. "We did everything a hospital would have done." It's a shitty line, trite and worthless. But it's also the truth, and it's often the only thing I can think of to say. I feel small and mean when I say it. Every single time. Even with somebody with no reasonable hope of resuscitation, the moment when I quit trying is hard for me. I don't feel responsible—I'm there only to make things better—but when I've poured all my effort and intelligence into someone's rapidly diminishing soul, it's hard to admit when the time comes to give up. One second the entire crew was frantically pounding and squeezing and pushing drugs, and the next second it was over. We stood up, stretched our backs, and started picking up the used needles and mounds of biohazard garbage we'd created. "We did everything a hospital would have done," I said again. Except that after we finish, all we can do is leave the body of your loved one lying on your own kitchen floor, stripped naked and stuck full of tubes.

If I'd thought in the beginning that I was immune to empathy for my patients, I was wrong, if only in degree. It's true that I never fall to my knees screaming at the injustice of the world. Nobody does. You'd never be able to keep going out on runs if every death affected you like it was your own father, if every tragic addict was your best friend. But living your life with a window that constantly overlooks human suffering can't help but affect you. I went out with Shona one night, and she asked how my last shift had been.

"Do you ever have one of those days when . . . ?"

"When what?" she asked.

"Never mind. It wouldn't make sense."

"So what? Just tell me."

"What I was going to say was, 'Do you ever have one of those days when everyone you meet at work ends up dying?'"

Shona laughed and hugged me. "Of course not," she said. "I don't think I'd have much of a future in the legal world."

Friends always ask me if I have nightmares about dead people I've seen. They ask if watching someone die haunts me. They ask me how I handle it. The truth is, I don't know how I handle it, how any of us do. You just put each call behind you and get ready for the next one. You tell yourself that the worst thing you can do is fail to work a miracle. The high of saving a life is so much dramatically higher than the low of losing a patient is low. I talk about my saves for weeks; I replay them in my mind. I feel like I own certain corners of the city, and every time I drive past a particular building, I think about how *that's* where I started a woman's heart again or *that's* where I opened the throat of a baby who would have died from his asthma without my help.

The dead don't affect me nearly as much as do the living, those people who exist on the fringes of death. Kids with rat bites. Heroin addicts who haven't left their pitch-black tenement hotel rooms for years. Diabetics so ill informed or apathetic about their condition that they have to undergo multiple amputations, slowly losing their limbs inch by inch as their circulation deteriorates. Sometimes I lecture people, tell them that they have to take care of themselves or they'll die. Mostly they ignore me; people need me to help them with what hurts *right now*, not tell them how to keep from hurting in some vague future that they can't imagine they'll ever live to see. Most pain is pain I can't cure, that doctors can't cure either, because it's chronic, because age and addiction and disease always win out over the elusive promise of lasting ease. Our presence as firefighters, the fact that we make an effort for people who are rarely taken care of, is often the only palliative that can be offered.

We go to nursing homes a lot, and in too many of them, failure is the acceptable default position. They feel like human warehouses, where people are stockpiled until they can be disposed of. When patients become too burdensome, the staff often calls 911, so that we'll take the patient to the hospital for a day of evaluation—and the nurses left behind get a few hours of respite. I've always hated nursing homes. When I was in eighth grade, my junior-high music teacher took us to sing at her grandmother's convalescent home. During the second chorus of "King of the Road," Grandma had a violent seizure and was wheeled away. We kept singing, though. She died ten minutes later, while we were singing "Rockin' Robin." Mrs. Lee didn't say a word on the drive home, and for once we didn't heckle her.

Now, just crossing the threshold into a board-and-care facility makes me feel empty and vaguely sick. Not just because of Mrs. Lee, but because we go there so often and the despair is always palpable. There is very little that we can do; the dominant condition is chronic loneliness and neglect, which attack the body as surely as any diagnosable disease does.

I saw an old woman once, sitting on the edge of her chair in a tidy, soulless room in an anonymous nursing home. She was so thin that her ribs threatened to break through the skin with each breath. Her eyes were wild, and she scanned the room frantically, searching for something she couldn't find. Everything in the room was in place, just as it is in every nursing home, from the plastic pitcher on the rolling bed stand to the heavy floral drapes and the permeating smells of disinfectant and atrophy.

She looked down quizzically at the hands in her lap, which were moving rapidly over each other as if of their own accord. The name on the door—Rose Levy—and the Jewish women's magazine on the bed stand made me feel immediately warm toward her. In my mind I treat all my patients the same, but I can't help it: Tribalism is a powerful thing, even for those who are connected to the tribe only by the weakest of threads. Pretty much all of the Hebrew and Yiddish I know is food- or insult-related, but occasionally I slip in a "Good *Shabbos*" just to see the surprise register on the faces of elderly Jewish patients who've probably never considered that their friendly neighborhood fireman might also be one of the chosen people. I wish I knew how to say "You're going to be just fine" in Yiddish.

I leaned down toward the woman in the chair. "Mrs. Levy? I'm from the fire department. Are you hurting anywhere?"

She looked up as if noticing me for the first time. Her eyes were wide and scared. "What do you mean? Why do you want to hurt me? What are you saying?" she asked. As usual, there wasn't a nurse anywhere, and I had no idea why we'd been called.

"Are you feeling any pain? Do you hurt?"

"I don't know why you'd say something like that. I don't know why you're asking about me." She pulled back from me as if she was afraid I might hit her.

"I'm asking because somebody was worried about you, Mrs. Levy. We want to take care of you."

She looked dubious and stole a glance at the three large firemen standing quietly behind me. "You're not going to hurt me?"

"I promise. We want to help you."

She considered me for a second more, then tilted her head to the side like a child. "Will you rub my arms? Will you hold my head? Everything hurts me. My arms, my arms . . ." She stretched her arms out to me, and, not knowing what to do, I took both her hands in mine and rubbed up and down her forearms. Her skin was brittle, like paper, with the thinnest trails of blue veins spiderwebbing up the backs of her hands.

"Will you wash my ears? Will you rub my back? Wash my ears, please, please. Nobody ever washes my ears. Won't you help me?"

"I'll help you, Mrs. Levy," I said. "I'll rub your arms." The ambulance was delayed for some reason. My crew saw I didn't need them and went outside to smoke cigarettes and stand in the sun. I sat next to Rose and rubbed her arms. Her head started a palsied nod, and her eyes closed. She stretched her arms my way and leaned in against me. I didn't say anything, just took off my latex gloves and rubbed her thin arms, up and down, up and down, wrist to elbow, elbow to wrist.

When the ambulance crew arrived, she opened her eyes, and for just a moment I know she saw me through her fog. Her head stopped nodding, and she patted my hand. "You're a good boy," she said. "You remind me of my son."

"I'm glad. Call back if you ever need me," I said, and then I went outside to join my crew in the sunshine.

Wake-up Call

There are no gentle good-mornings at the firehouse; I wake up most days as if I've been kicked. I never remember tugging on my socks or stepping into my boots, pulling the suspenders up and over my shoulders. But my heart is awake immediately, pounding at the inner wall of my chest, running leash-free until my brain catches up and asks it to slow the hell down. After a run I often lie in bed for hours, getting furious as I hear the others drift back down easily. Shift change in the morning is at eight, and I always try to sleep right up until the minute I'm relieved. But more often than not there is a call at six-thirty, and when we get back with the sun breaking the clouds, there is no point in even attempting to lie down again. Those are the worst—the ten-minute false alarms that rob you of an hour and a half of much-needed rest.

After one particularly rough night at Station One, I came stumbling blearily down the stairs just before the change of shift. Dominique and the other hookers had been jostling for position under my window all night long, and their catcalls had kept me up. Around 5:00 A.M. I'd heard louder screaming than usual; one customer had been surprised to find out that Dominique was a man and was trying to get away. Dominique, for her part, wasn't about to get out of the guy's Mercedes until she'd been paid, service or no service. In the midst of the standoff, Tom shuffled downstairs with a sigh of resignation and brokered a truce: Dominique got half her normal fee, the guy got to leave, and nobody had to see the cops.

I went to the rig to pull off my gear so the new shift could get set. My turnout coat was draped over the end of the ax where it lay in its bracket alongside my seat. I pulled the coat down from the ready position, wrapped it with the strap from the helmet and the leather utility belt I wear, and was preparing to throw the bundle into my locker when Jack Alvarez cornered me.

"How come you don't eat with the guys?" He stared at me and, as usual, scowled. I always got the feeling that Alvarez had made up his mind about me on the first day we met, and that his opinion wasn't particularly charitable. He was a big, bald Hispanic guy with a wiry mustache and that perpetual frown. I tried to think of some meaningless witty response, but I realized that I had absolutely no idea what he was talking about.

"I heard that you bring bag lunches, Harpo. Don't you want to eat with the crew?" Alvarez asked. He was the self-appointed big dog on his shift and saw one of his duties as the bestowing of nicknames, most of them designed to reveal a colleague's weakest feature of attitude or appearance: Snacks. The Missing Link. The Spleen (he's a useless body part). Mustache on a Stick. Chardonnay (he's a fine whiner). Ratboy. While firehouse nicknames are often more than a little mean-spirited, they also convey a sense of belonging; if you're getting hassled, it means that people care enough to notice you. One firefighter I know desperately wanted a nickname and slyly tried to assign himself a tough-sounding one, going so far as to write it in the collars of his shirts and leave pieces of paper with his new nickname lying around conspicuously. His ruse was of course immediately discovered, so now we call him Tinkerbell or Daffodil—anything to deflate his misguided ego.

Jack had taken to calling me "Harpo" for my odd curly hair, which makes me look more like a circus clown than a grizzled smoke-eater. When your last name is Unger, any nickname other than Felix is a welcome relief. Plus, I liked it better than Prozac.

"You just want to be by yourself because you don't like us, is that it?" Alvarez pressed.

The desire to be alone is perhaps the greatest of firehouse sins. In a job that often revolves around boredom, it's imperative that there be enough targets to keep life interesting. People who hide out in their bunks are thought to be "one-way" guys, firefighters interested only in themselves. Somebody who disappears at home will disappear on the fireground, will find a dark, smoky room to hide out in while the rest of the team are working their asses off. Since the kitchen is the epicenter of firehouse culture, any attempt to absent oneself from the table is a snub. When one old officer from Station Three finally turned in his badge, they didn't know what to serve at his retirement dinner because nobody knew what he liked to eat. Everyone said he was pleasant enough, never actively pissed anyone off, but he'd spent every lunch and dinner up in his room with food he'd brought from home, which made people distrust him no matter how nice or competent he might have been.

"You go to Alameda if you want to be alone. This is Oakland," Alvarez said. In Alameda, the small, wealthy island just off Oakland's coast, the firefighters rarely ate together. Instead they brought in leftovers from home or swung the engine by a sandwich shop between runs. The distinction was clear: Big cities that fight big fires need to eat big meals together.

I saw what he was getting at, though. The constant steak-and-Caesar nights were starting to take their toll: I'd already let my belt out one notch, and I wanted to stop the damage before it got any worse. So for a few weeks I'd been picking around the edges of the meals, often bringing in fruit or yogurt from home. I loved meals—the lying, the gossip, the jousting—but the California fitness craze seemed to have passed by the firehouse without leaving so much as a carrot stick in its wake. "Just trying to get healthy, you know," I said.

Alvarez glowered, unconvinced. "You like to be by yourself, huh? Tom told me that for your last job you just sat out in the snow by yourself and watched birds fuck." His style of joking was indistinguishable from his style of getting angry—he'd perfected the dirty look for most forms of communication. I decided to assume he was kidding. It seemed like the more pleasant alternative.

"Alvarez," I said, "sometimes watching birds fuck is a hell of a lot more interesting than listening to all you C-shifters whine at each other." I was determined not to give him any edge, to make light of everything he and the rest of his crew could throw my way. It was a never-ending process. And it involved a lot of swearing.

He softened a little, scratched his head vigorously with both hands as if he were trying to physically wring sleepiness out of his head. "I guess we eat too good for you tree-hugging nature weenies. Next time you work on our shift, I'll make you some fucking nuts and wood chips so you can eat with us." I'd read him right. He'd been kidding but sending a message at the same time, that eating cottage cheese was an affront to the solidarity of the crew.

"Watch your own damn plate, fat boy," I said, eyeing his stomach. "You could use it."

He pulled up his shirt, thrust out his midsection like he was pregnant, and rubbed his hands over the Buddha belly. "Fuck you, Harpo," he said, breaking into a crooked, toothy smile. "Get off my truck now."

"Fuck you, too, Alvarez. Be safe today."

I wandered into the kitchen and had a cookie for breakfast. Another half-malignant, mostly harmless attack defused, and it was time to go home for two days of peace, quiet, and uncontested celery stalks.

. . .

For my first year in the department, the C shift at Station One had been my biggest bugbear. Every working morning as I was getting ready to go home, they'd be coming on shift, raring with energy and looking for something to break. I'd often wake up in my bunk to the sound of weights crashing or pots being slammed onto the stove in a violent display of oatmeal cookery. They would light into one another immediately, yelling curses and wrestling. They'd been penned up at home for two days, playing nice with the wives and kids, and now that they were back at work, they had to exercise some sort of caveman inner nature.

My shift, the B shift, was made up of calm old-timers and eager new kids like myself. The C shift was dominated by men with about ten years in: too old to be sincere or polite, too young to have learned anything about moderation. They went through cycles of fraternity and petulance. One month they'd go on hunting trips and announce their solidarity with beery embraces. The next they'd hardly be on speaking terms, sulking around the firehouse, casting bitter glances at each other. When a spot would come open after a regular guy was injured or promoted, they'd try to woo somebody from the opposite end of the city or else draw in some like-minded new kid. Almost immediately the new member would be sucked up into the web. "The C-shifters eat their young," Tom would say, chuckling. "Just look at 'em; they're like a bunch of rats locked up in a tiny box together." You could tell that Tom liked the C shift. He liked their aggressiveness and what was in the end an unshakable commitment to each other and to the job. I imagine he saw in them the way it used to be for him before all his buddies either retired or moved up to cushy assignments in the hills.

Jack Alvarez was the heart of the C shift, and he tried to impose his rule with varying degrees of success. He'd been an ironworker before he was a fireman, and somebody on his shift had written IRON WORKER UNION CARDS on the wall over the toilet-paper dispenser. Alvarez's eyes shifted constantly, unless he'd decided that he needed to stare someone down, in which case he could spend an entire meal drilling holes into an uncomfortable new kid who was only trying to keep a low profile and eat his dinner. It was rumored that at home Alvarez slept all day. But at work he never stopped moving, vibrating in place during roll call or while standing in front of the stove. He'd stay up until four in the morning, tinkering with equipment or building new things for the firehouse.

Like most truck crews, the C shift considered themselves the best in the city. And also like most truck crews, they had a credible claim. Truck One is located at Sixteenth and Martin Luther King, right in the heart of downtown. And downtown is not an easy place to fight a fire. Sizing up a burning tenement or high-rise requires a set of skills wholly different from those needed to kick ass at a small bungalow fire. Alvarez and his crew were expert at spotting the rig perfectly and threading the hundred-foot aerial ladder between streetlamps and high-tension wires. They knew where to find the standpipes, the alarm panels, the hidden back stairwells, and the street elevators that led to the basements. They were comfortable on the pitched roofs of churches and on the rickety, rusted fire escapes that clung to the sides of Bowery-style motels. The officers on the shift were easygoing and pragmatic, experienced enough to trust their crew and smart enough to know when a little guidance was necessary.

Additionally, Station One was home to Oakland's Heavy Rescue program and to the rescue rig. The "rescue" was shaped like a bread truck topped by a crane. The rig's engine obviously hadn't been built for the mass of tools that were loaded into it, and on the highway it was a gutless wonder. I always found it embarrassing to be running toward a call with full lights and sirens and watch some blue-haired grandma passing us in the slow lane. Alvarez was the acknowledged Rescue

Guru. The tools in the rescue rig were for specialized situations: building collapses, train wrecks, underground cave-ins, high-wire rappelling jobs. The guys on the C shift knew what they were doing when it came to these odd jobs. They were all certified as instructors in some aspect of mayhem mitigation, and Alvarez and the others would often be out in the back parking lot late at night stringing up lowering systems or lifting cars with airbags and hydraulic tools. They'd worked on the collapsed Cypress Structure section of freeway that had killed forty-two people during Oakland's earthquake in '89. They'd been to Northridge for Los Angeles's deadly quake, to Napa for the floods, and to a thousand car wrecks and minor disasters in between.

One thing the C shift did not do, however, was make new kids feel at ease. In the mornings I often tried to avoid them. But with vacation coverage, overtime, and the practice of firefighters on different shifts trading days around, it seemed that Jack was always at work on my shift. Occasionally I'd work a trade or an overtime shift and end up spending twenty-four hours in the heart of his corral. Those days I tried to stay silent and busy. I'd go over all the tools again and again, help out in the kitchen, do an extra share of housework. When Alvarez checked out the rescue in the morning, I'd be by his side, watching, learning, trying to ask intelligent questions.

It was clear that they didn't know what to make of me. I didn't suck up or try too hard to fit in. I wasn't dismissive or haughty, but I tried not to care too much what they thought of me either. I'd made a commitment to myself that I wasn't going to lose my identity to the pack mentality of "what a firefighter should be like." I wanted to learn the job and wanted to be good at it. I wanted people to hear my name in passing and say, "Unger? Yeah, he's a good fireman." But I'd resolved that I wouldn't go to extremes just to prove how far backward I could bend. The C shift, like firefighters at most truck houses, liked guys who would perform new-kid acrobatics, like diving over tables to be the first to answer the phone or staying up until dawn polishing brass fittings.

When I worked with them, they'd hit me over my background, my appearance, my cluelessness—all the usual sort of meaningless probie harassment. I'd just bob my head and smile and let them work me over until they got tired of it. I'm not quiet—not by any means. Now I sling insults and grouse along with the rest, but back then I did my best to keep quiet and just watch. They'd try to get a rise out of me for a little while, then give up and turn back toward each other, where they knew exactly what to say in order to stir the pot.

One day when I was working on the C shift, we got our routine ringing-alarm call to the St. Mark's Hotel. The roll-up door on the firehouse didn't have an automatic closer, so the driver would pull forward until he cleared the door. The junior member (me) would hit the button and then run to catch the truck before it went off. That day when I closed the door, they just kept right on driving, leaving me standing there all alone. I saw Jack give me a "so long, sucker" wave from the tiller bucket before they turned the corner.

Fortunately, I was able to slip under the door and back inside before it closed and locked me out. I walked to the kitchen and sat back down to the meal that had been interrupted. I knew they wanted me to feel angry, disappointed, tested, but all I could muster was an overwhelming fatigue. I thought about getting worked up, but the truth was, I wasn't missing anything more interesting than just another ringing alarm, and I couldn't make myself care. I thought about how the guys on the truck must have been chortling like third-graders who'd pulled up the rope ladder to their secret tree house. It did feel odd sitting at the big table all alone; I'd never been in the kitchen when it was quiet. I covered all the plates with tinfoil to keep the food warm.

Predictably, in just a few minutes, I heard the diesel rumble that announced their return. As always, they'd reset the alarm with no difficulty and cleared the scene quickly. They walked back into the kitchen together and saw me sitting with a fork in my hand. Alvarez stared at me in mock amazement, then thrust out his chin and said, "That's some new kid right there. Missing a run to eat dinner while the rest of the crew's out working. I tell ya what, guys, this kid's *fucked*." He eyed me, gauging my response. The rest of the crew was rolling in loudly, sitting down at the table with rough chair scrapes and the clang of silverware. When Alvarez shot at me, everyone went quiet to hear what I would do. They were hoping for a mewling apology, a new-kid plea for forgiveness that could serve as the jumping-off point for a few minutes of ridicule.

"I don't mind," I said midbite, barely looking up from my plate. "It's just five less minutes I have to spend with you guys."

For a quick second, everybody was silent. Alvarez started to sputter an objection, but before he could, the rest of the crew began roaring with laughter. "He got you, buddy!" somebody shouted. Alvarez was momentarily speechless—a rare and small victory on my part—as the focus of everybody's laughter shifted away from me. For the briefest of instants, I'd made a fool of him. Quickly enough he realized that I'd won the moment, so he shifted tactics to run with the joke and make it his own.

"Unger doesn't like us—did you hear that? He hates us! Damn cake eater doesn't like to hang out with the *workers*." They all joined in, but it was clear that my comment had changed the air. Now I was talking their language, calling them pieces of shit. They could understand that, and they liked it. It wasn't that I came to match them in rowdiness or aggression—I never have. But I'd made it clear that I had a personality, that I wasn't intimidated by their swaggering. From then on, whenever I worked on the C shift, Alvarez would say, "Oh boy, here comes that guy who doesn't like us. Are you going to be able to tolerate our bullshit today, or should we all just leave you alone now?" The tone had changed from the day they left me standing in the doorway watching the truck make the corner and go out of sight. I'd never be a C-shift insider—or want to be—but they'd started to throw a personality on me.

Whether they liked me or not, I'd become a person instead of just a faceless new kid.

I'm not sure why I picked that moment to fire back for the first time. It's funny how relationships can turn on little things, how a single line muttered half under my breath without thought could cause an entire shift's worth of people to reevaluate what they thought of me. More than anything else, finding your way in the firehouse is about learning where you fit in. Not everybody has to be loud and obnoxious—though it's certainly more entertaining when at least a few people are. But it's good to prove that you can take a hit and throw one back, that you're unflappable, that you were your own man before the fire department and that you'll go on being your own man long after you're too old to swing an ax. Firefighters prize ability and work ethic, and in the service of those, a little eccentricity is easily tolerated, maybe even encouraged.

In the grand afterglow of my well-received fuck-you to the C shift, the Klaxon rang and the printer spit out a call for an "Extrication/Rescue." Station One sits just astride the boomerang curve of a freeway, so close that through the window of the bunk room I could hear the gears of the big rigs wind up for a short incline late at night. Within a mile of "our" on-ramp, four freeways come together and cross over. It's a good place for wrecks.

Whenever I grab the printout or check the dispatch screen, I always look first at the line, titled "Nature of Call." That's where dispatch has reduced the reason for our run to a word or two: "Bleed/Hemorr"; "SOB"—that's shortness of breath, but with some of our regular customers, it might as well stand for something else; "Assault/Stabbing/Gunshot Wound"; "Abd Pain"—usually known as "Abominable Pain"; and the ever-popular "Unk. Illness." That last one—"Uncle Illness" to us—is a catchall for anything that doesn't fit neatly into any of the other categories. Nobody calls 911 and screams into the phone, "I have an unknown problem! Please send help!"

While the stated nature of the call is often wrong—I once went to a "Bleeding" call that turned out to be a newborn still dangling from its umbilical cord—it gives my mind a head start. "Extrication/Rescue" calls are good ones. We're always happy to see those words on the nature line rather than yet another "Unconscious/Fainting." A car wreck offers the promise of an acute trauma, something that quick action can actually fix, rather than a chronic malaise. It's also no small benefit that car wrecks are outside—no tenement stairs to carry someone down, no rancid hallways to negotiate. Just the good clean smells of spilled gasoline and torn upholstery leather. Plus there's the kid thing. We're like little boys who get excited by big crashes, heavy machinery, and shattered glass. Stomach pain accompanied by persistent diarrhea doesn't have the same appeal.

Car wrecks are the most dangerous things we do. In the drill tower, Gold had ordered us to look both ways and yell "Stepping down!" every time we stepped off the rig. While it had felt a little *Sesame Street*-ish at the time—we were, after all, grown men, heroes no less—it made scary sense after a few shifts of observing Oakland's maniacal motorists. Flashing lights and flares mean nothing to the majority of them. With their attention divided by stereos, cell phones, and fast food, it is foolish to think that the safety of roadside emergency workers registers even a blip on the radar screens of most drivers. I'd feel so much more foolish dying on the side of the freeway than in a house fire.

Which is just to say that when Alvarez pulled up alongside the mangled car, I climbed awkwardly over the engine shroud so I could get off the truck on the side without traffic. A black luxury car was sitting at the end of a long skid mark. The air still smelled of burned rubber and expelled air. A smear of fluids—reds, greens, blacks—had trailed out as the car crashed, and globs of motor oil and antifreeze made greasy rainbows in the puddles at the edge of the freeway. The car had spun around, maybe more than once, then glanced off another car and into a signpost. While Jack was setting the chock blocks under the wheel of the rig so it wouldn't roll away, I went straight for the Hurst tool.

The Hurst tool is what you would call the Jaws of Life if you didn't know that it's correctly called the Hurst tool. I've never heard a fire-fighter use that stronger, more descriptive phrase. Maybe it's too self-consciously heroic. Maybe it puts too much emphasis on the tool and not enough on the operator. In any case, and with any name, it is an impressive piece of equipment. Powered by a rolling generator, the Hurst tool is really a collection of tools that can attach to the end of a long hydraulic line. Cutters, rams, and spreaders. All designed to tear the hell out of a car. I took the heavy spreaders in one hand, and with the other I helped another firefighter carry the generator.

The driver was still sitting in her seat, hands on the wheel and her eyes on the road, as if she could just drive away. She was small and pretty, with long black braids and a thick wool sweater. I felt bad for thinking she was pretty; she was probably scared for her life, and all I could think about was that in different circumstances she was someone I would want to kiss. I poked my head inside the window, introduced myself, and immediately felt stupid, as if I were some sort of rescuer maître d'. She was calm until I talked to her. The steering wheel was in her lap, pressing deep grooves into her thighs. The bottom of the dash-board was pinning her legs. The car must have hit something head-on, and the weight of the engine block had been transferred back into the passenger compartment. The door was crumpled closed, the hood buck-led in a steep pyramid, and everything that had a minute ago been a safe arm's-length in front of her was now trapping her in her seat. She was tilted forward, the two front wheels having blown out; she was locked tight by the seat belt and gasping a little as it pressed against her chest and throat. When I put my hand on her shoulder, her lip started to tremble, but she never stopped looking ahead.

Alvarez hustled me out of the way with a bump. He was holding a carryall, a thick piece of canvas that we use to haul water-soaked fire debris out to the curb. He opened the passenger-side door and knelt inside. "I'm gonna put this over you," he said to the woman, "just to keep you safe." He didn't want broken glass to fall on her or strips of metal to lash out when they popped from the strain of the Hurst. She didn't complain as he shrouded her in darkness with the ash-reeking cloth. As we went to work, she sat upright and still, like a kidnap victim under a blanket.

"Hand me the spreaders, babe," Alvarez said to me. He called everybody "babe"; I wondered what he called his wife. The generator was running, and the tool was humming in my hand. I hadn't used the Hurst yet at a real job. I'd popped a couple of doors here and there after the victims were out and torn the hell out of a bunch of wrecked cop cars down at the drill tower, but I hadn't used the tool yet when it actually mattered. Handing it over was not an option.

"Give it up, babe. Let's go." Alvarez had his hands outstretched and was raising his eyebrows at me.

"I got it," I said, shouldering past him.

Alvarez wasn't going to win. He was itching for the tool, but he wouldn't have respected me if I'd deferred to his experience and given it up. Of course, everybody should be rescued by the ablest rescuer, but if only the ablest rescuer ever gets to do the job, then nobody will ever be trained to take his place. The credo of the new kid is never to give up your tool. If somebody says, "Hand me that ax," the only proper response is, "Show me what you need done." Somebody who gives up his tool is a lump, a jake, a coward. I didn't have a clue what I needed to do, but I wasn't going to let that slow me down.

"Well, get the fuck started, then," Alvarez said.

He took the Halligan forcible-entry tool in his hand and worked the flat end of the blade between the post and the door, in the thin gap just beside the keyhole. The metal sheared back easily, opening up a jagged tear. I had the tool resting on my thigh, and when Alvarez moved aside, I heaved my whole body toward the opening. The spreader looks like the front end of a giant scissors; once the points were in place, a touch of my thumb on the control switch started the tips moving powerfully apart. The metal screeched and tore, popped but didn't give.

"Pull it out, Harpo," Alvarez said, loud enough to be heard over the traffic, but gently, instructing me, "You've got to reset it farther down." I'd bent the flimsy part of the door aside but hadn't gotten a good hold on anything solid. I pulled back, closed the spreaders, and then lurched back in for another shot. I leaned against the car to brace myself.

"Get on the outside. You'll get pinched." Alvarez motioned me to step out from between the tool and the car so that I wouldn't be caught between. He leaned his back against mine, and I was glad to have something solid and safe to brace against. I reset the points and thumbed the controls. The metal spread quietly for a second, groaned, then popped with a bang. The door tore open, and I fell back against Alvarez's weight. He stepped away from me and leaned heavily against the door, stretching the springs backward until the door lay flat against the front of the car.

"There," I said, looking inside at her. "We've got the door off. We'll have you out in no time." The tarp nodded. I felt down into the wheel well with my hand. Groping around, I could feel only hardness—nothing that felt like legs. Crumpled steel and plastic.

"Can you feel my hand?" I asked when I touched something that felt almost human.

"No," she said, muffled. I felt lousy for having to ask; she was smart enough to know that not being able to feel her legs was a bad sign. I reached down again, running my hands along her legs until I could feel a bone protruding at an odd angle just below her knee. That is one of the things that scares me the most: broken glass, torn gloves, the commingling of a stranger's blood with my own. I told myself that she looked safe. I tell myself that about everyone.

Sometime while we were popping the door, the ambulance arrived.

One of the paramedics crawled in beside the patient and was working on starting an IV. The patient's right arm stuck out from underneath the carryall. The medic was rubbing the arm up and down with an alcohol prep pad and walking her gloved fingers over the skin, trying to find a suitable vein to puncture. I patted the patient's other hand and talked to her, leaning close to her ear to be heard over the traffic and the Hurst generator.

"They're just going to start an IV, put a little needle in your arm to give you some fluid."

"Okay," she said, her voice entirely flat. She was trembling a little.

"You're lucky, too. This is our best crew working tonight, and they'll have you out in no time." I kept making words just for the sake of talking, of keeping her calm.

"Okay," she said again. I didn't know if she wanted to be alone, but I knew I wouldn't if I were in her spot, so I kept close to her. When she spoke, her voice was soft and low, with a light Caribbean singsong that I liked. Cars and trucks flew by, and someone yelled out of a window, but whatever he'd said was lost in the speed.

"Let's peel the roof," Alvarez said, slapping his hands together. "Out of the way, ambulance." The medic hesitated, then withdrew. Alvarez had the cutters in his hand, malevolent-looking metal lobster claws. He went to work on the support posts. He cleared out broken window glass with his gloved hands, then thrust the cutters into place on the posts and pressed the button to close them. The front ones went first— left, then right—yielding easily under the hydraulics. Alvarez worked quickly and with assurance, cutting the posts where the seat belts attach. Simultaneously, and without saying anything, Alvarez and another firefighter brought their fists down on opposite sides of the car roof, making a deep dent. Using the dent as a fulcrum, they peeled the roof back like someone opening a sardine can, and suddenly we were working on a convertible.

"We've got the roof off," I said, as if that were something to be excited about. She peeked out from behind the canvas veil and let it fold into her lap. The sky was beautiful. Reds fading to black over the water, a chubby half-moon in the sky.

"Am I going to be all right?" she asked, looking up at me with big, clear eyes. She was breathtakingly pretty.

"We'll get you out."

We'd dismantled the car from around her, but she was still firmly in place. I'd been hoping that we could take her up and out through where the roof had been, but for Alvarez, peeling the roof had been only the first step. "Get the chains and the ram, and we'll roll the dash," he said. I didn't want to leave her, but sitting with her and holding her hand wasn't getting her any closer to the hospital. I went scurrying back toward the truck, forgetting where I was. A dirty wind hit my face as a Budweiser truck whistled past me, still going full-out despite the row of fire trucks and highway-patrol cars all lined up in the speed lane with their flashers on.

When I came back, Alvarez had the spreaders lying on the crumpled hood of the car, the jaws opened as wide as they could stretch. He snapped the heavy chains out of my hands and dropped a loop around the column of the steering wheel. I wrapped another chain along the thick part of the bumper. The chain ends met in the middle of the hood, hooking onto the outstretched arms of the Hurst spreader. The idea was to use the pulling force of the tool to roll the dashboard up and away from the woman's legs. I knelt back in the car beside her.

Alvarez readjusted the heavy tools after each pull and tightened the chains methodically. The medics had an IV line in, and now they had nothing to do but fuss with their gurney, raising and lowering it, messing with the straps, trying to look busy. The highway cops were pacing off distances in their high boots and jotting down notes on little pads. The traffic was moving quickly, almost recklessly it seemed; standing on the side of a freeway is unsettling. You never feel how dramatic sixty miles an hour is when you're sitting inside a car. The patient—she said her name was Lina when I finally thought to ask—leaned her head back on the seat and stared upward, past it all. The beads at the ends of her braids chattered musically with the wind. She never cried or closed her eyes, never asked me what we were doing, but she held my hand tightly. I stopped explaining things to her. She just wanted us to make it better, to make sure she got to someplace safe. She didn't care about how we were going to do it. Sometimes I wonder if I ever truly do make things better in a way that's real and lasting or if I'm just a tool that's used to pick through the wreckage. I see a patient, and then she's gone. I offer a little bit of hope, a few soft, comforting lies, and then tell myself it doesn't matter much and that I'll do it all again the next day. Once the patients are in the ambulance, we never see them again.

The freeway is actually beautiful in the right spots. From where I stood, I could see the far end of the Bay Bridge, outlined in lights as it curved toward San Francisco. Behind me in the dusk, I saw the Oakland skyline, the lights of Lake Merritt. Plus, the freeway is clean—it's too windy for trash to stick—and there's something appealing about the futuristic layered maze of ramps and merges and flyovers. So I knelt in the backseat with a hand on Lina's shoulder and felt useless.

"Be ready to move fast when the dash rolls," the medic whispered into my ear. "That's when they always bleed out. It's like they wait until you're ready to save them, and then they just die on you." I'd heard about this before but never seen it. Sometimes the compression of the dashboard against a person's legs can act as a sort of tourniquet, preventing any loss of blood as long as the pressure stays on. When the dash rolls off, the pressure is released, and a seemingly stable person can rapidly bleed to death from injuries to the lower body. A trickle of blood ran from a tiny cut on her forehead, but beyond that she looked unharmed, as if she were waiting in her car at a traffic light.

"We're close," I said. "We'll have you out in a second." She squeezed my hand harder, and I couldn't decide if I wanted to think of her as a rescue problem or as a real person who was communicating her fear by squeezing my fingers. I took off my thick glove and put my bare hand back in hers. She let her hand lie in mine, unmoving. I concentrated on the details and the heavy sounds of chains and tools, telling myself I should be learning by watching Alvarez work. But all I could see of him was his upper teeth biting the corner of his lower lip.

The dash came away with a groaning that sounded more like exhaustion than defeat. I had expected that Alvarez's procedure would lift the dash off like a banana peel. Instead the change was barely perceptible, just a shift of a few inches that meant her freedom. The steering wheel bent upward, then caught, then folded back on itself toward the front of the car. For the first time, I could see her legs. They were stained dark black through her tan pants. She might have smiled at me. We put her on a backboard and strapped her down. Somebody said, "On my count," and we moved the backboard to the stretcher. Somebody said, "One, two, three," and the stretcher went into the ambulance. Alvarez said, "Nice job, Harpo. Let's get back in service." The ambulance drove away, took an exit, and as we were cleaning up, I could see them on a lower level coming back toward us on their way to the trauma center.

I never told her that she'd be all right, but I let her think that. After a patient leaves in the ambulance, she's gone and we never find out what happens to her. I remembered how pretty she was and how she'd closed her eyes and seemed to deflate as soon as we got her onto the gurney.

Back at the firehouse, I pulled the generator and the Hurst tool out of their cabinets and worked on getting them ready for the next call. I checked the hydraulic fluid and fuel reservoirs, made sure that the tips of the spreaders were positioned just right, wiped the grease and road dust off the chains with a rag. I wondered if anybody else felt as empty as I did right then. We'd saved somebody for a minute, sure, but then she was out of the car and out of our lives.

Jack Alvarez came up alongside me, cup of coffee in his hand. "Harpo—I think you liked that girl," he said, gesturing at me with his cup.

"I don't think so. I was just trying to help her out," I said. I didn't look up from where I was fumbling with the cap for the hydraulic oil reservoir on the Hurst.

"Now, don't you give me that," Alvarez said, puffing up his chest in a theatrical attempt to seem threatening. "I saw you checking her out. You must think I'm dumber than a bag of hair if you think I couldn't tell you liked her."

"You got it, Alvarez. I love a woman in crisis." I rocked back on my heels and looked up at him for a second, then bent down again to my work. I was grateful for his banter. It was rare for Alvarez to be silent, so I kept working and waited for the inevitable next shot. He was the kind of guy who was so unrelentingly negative that you couldn't possibly take him seriously. It was probably just easier for him to enter a conversation with a predetermined point of view.

"Harpo, is there a single person in your family who's ever worked a blue-collar job?" Alvarez asked.

"Well," I said, "my grandfather used to wear a blue blazer sometimes. Does that count?"

Alvarez laughed and exhaled. His game lay in trying to find out whom he could intimidate; when his jabs couldn't find any purchase on me, he softened. "You're a funny fucker, you know that? I know I don't understand you, but you're funny. Your parents must be so depressed."

"Why's that?" I asked.

"Because they blew a hundred thousand dollars on sending you to college, and you're still just a fucking fireman. I'll bet your mom cries every night—'My baby's a fireman! Why couldn't he be a doctor like all the other little rich kids?'" I thought of my grandfather beaming at my college graduation and of the way he'd asked me years later if being a fireman would be enough for me. It felt like enough.

"The little professor. You're the kid we used to throw rocks at in grade school. We used to chase you and the other propeller heads around with sticks. And now I have to work with you? That's fucked up." Alvarez was just having fun now, no harm intended, no offense taken; telling someone how worthless he was is just a fireman's way of passing the time.

Alvarez looked at me and cocked his head to the side, as if suddenly struck with an idea. "Professor . . . you know what?"

"No, Alvarez, I don't. Tell me what," I said.

"I'll bet you're writing a book about us. That's why you're here, isn't it, Harpo?"

"Who would the bad guy be?" I answered. "Are you auditioning for the part?"

"No, it wouldn't be that kind of book. It'd be a study or something." He held up an imaginary monocle to his eye. "'The fire people are a strange and backward tribe. During my years living among them, I observed their many horrible customs.'" He broke down laughing, rubbing the back of his bald head. "I'm right, aren't I?"

"That's not a bad idea, Alvarez," I said. "I'm going to blow this thing wide open. Thanks for the advice. I'll make sure you get the first copy."

Vice Squad

All flophouses look the same. Darkness above all else. Light is a
luxury that junkies can't afford. Old blankets are tacked over
the windows until they rot and get replaced by newspapers.
Mattresses lie on the floor, never on frames, several to a room. All the
amenities of life are missing: sheets, books, tables, running water. The
floors are invisible, too, covered with garbage, the meager remnants of
fast-food meals and chili eaten unheated from a can with the lid still
partially attached, a hazard underfoot.

There's a smell also. A smell so heavy it's tangible, coating every
surface, infusing every crevice. The smell mingles with the darkness,
and the dead air from inside is like an assault when you open the door.
A flophouse is impossible to miss; you know it the second the rig pulls
up in front. This one was over on Wood Street, just across the 880 freeway, but a world away from the bustle of downtown. Over where the
streetlights don't work and the potholes don't get filled and it always
takes the cops a few extra minutes to arrive when you need them. You

can tell you're in a really bad neighborhood if the cops put up street
signs that say DRUG-FREE ZONE.

"Home, sweet home," Billy said, pulling the rig over to the curb.
"Harpo, you could buy this place and walk to work. It'd be like a dream
come true." He reached into his pocket and threw a couple of quarters
at me. "I'll even lend you the down payment. Knock yourself out."

We grabbed the medical bags and walked to the house. Captain
Powell stood against the outside wall, poised between the doorjamb
and the first window—a habit, a police tale, a myth that if you stand to
the side of the door you won't be shot right through it. It's like any old
firefighter myth: Nobody's actually heard of its ever happening, but you
wouldn't want to ignore an established piece of folk wisdom and be the
first unlucky bastard to suffer the consequences. Standing to the side
while ringing the bell has become such an ingrained habit that I notice
myself doing it wherever I am, as if my friend Jeff might be lurking behind his door with a shotgun when I go over to his house to watch the
A's on TV.

Powell reached his hand across, knocking with the heavy butt of his
black Mag flashlight. "Fire Department. Anybody call the fire department?" A weak, choking voice came from within, and the door opened.
A woman met us at the door, turned without a word, and led us to a
back room. She was shriveled and toothless, a storybook hag with
pockmarked skin and sunken eyes. It was impossible to tell how old she
was; heroin corrupts every natural process. "What's the problem?" I
asked. She shot a hand up into the air and waggled it around without
saying anything.

On a mattress in a back corner sat a man with his knees drawn up to
his chest and his head dangling limp at an odd angle. His mouth hung
open. There were bits of food caked at the corners of his lips, and he
didn't move when the beam of the flashlight hit his eyes.

"What's going on?" I asked again. He didn't respond, so I squatted next to him—squatted because you don't want to put your knee down where there might be a needle. I rubbed two knuckles hard against his sternum.

He woke up from the pain of the rub, shook his head, and slapped himself in the face a few times to bring himself fully awake. "Are you hurting anywhere?" I asked him.

"That bitch," he said, after a coughing spell. "I don't know why she called you. I shot up with a bunch of bleach by accident. It's not such a fucking big deal." He pulled his filthy undershirt to the side and showed me an area on his upper chest that was red and blotchy, sprayed with tiny red bumps.

"I was popping," he said, referring to the practice of "skin-popping," in which addicts with no decent veins left shoot themselves just between the layers of skin and fat in hopes of getting the drug to absorb through the meat of their bodies.

"I always rinse my needles 'cause I don't want to get sick, but I forgot to flush this one. I thought that was water I was cutting the shit with," he laughed. "Guess it was still the bleach. Shit, I've done it before, and I'm still here." I turned away as he coughed again, a dry, tubercular rasp.

There was nothing to do for him. An ambulance was on its way. Billy, who usually held back on medical calls, stepped forward from the doorway where he'd been standing. He cleared a spot on the floor with the tip of his boot and went down on one knee. "How long have you been using?" Billy asked quietly, without any of his usual sarcasm. The man looked up at Billy sharply. He hadn't been expecting that question—it wasn't something people ever cared about, wasn't relevant to his emergency of the moment.

"Have you ever tried to quit?" Billy asked.

"Why you sweating me? I quit every fucking day."

"Don't quit quitting, then," Billy said, his ungloved hand resting easy on the man's shoulder. "I know where you've been. It'll be the hardest thing you've ever done in your life, but you can do it. You'll get it for sure. Don't give up."

Suddenly the man was crying, silent sobs that he buried in his shoulder. The woman who had met us at the door looked on. Her skinny limbs twisted strangely, seeming to bend backward at every joint. Her face was hard and unreadable.

The ambulance crew knocked and entered. Billy stood up, said, "Good luck," to nobody in particular, and walked back through the house to the front door. Pulling his seat belt across after he stepped into the engine, he said to Captain Powell, "He'll never quit. He's too far gone already. That guy's dead."

Billy backed the rig onto the cement apron of the firehouse. He stepped down from the driver's seat, uncoiled the fuel hose, and flicked on the diesel pump. As I filled out the fuel-report paperwork, I tried to figure out how to ask Billy what he'd been talking about back in the flophouse.

"Were you pulling that guy's chain to make him feel better back there?" I asked.

"Shit. That was me ten years ago," he answered, scratching with his nail at a bit of road tar stuck to the side of the rig. "We used to go on calls for drug overdoses, and all I would be thinking about was how I could steal whatever was left in the syringe that was hanging out of the guy's arm. Hell, if it knocked him out, it had to be good shit."

The story didn't fit with what I knew of Billy. He'd always been re-lentlessly, almost eerily, upbeat. Despite his irony and his caustic wit, his goal in life seemed to be helping other people. Every Sunday he would drive the engine to Children's Hospital and spend the morning on the burn ward, passing out plastic fire hats and making all the kids forget why they were there. He was like a whirlwind, flying around the city from grade school to baseball diamond to convalescent home, always smiling, helpful, happy to be alive. I couldn't picture him as an addict.

"I used to get so fucked up," he continued, "that I'd miss the pole entirely, just fall through the hole and hit the ground." He stopped for a moment and then laughed. "I tell you—I was a hell of a house guy when I was strung out though. I'd be down in the kitchen at three in the morning polishing the light fixtures and dusting the tops of the doorjambs."

"Did you ever come up dirty on the drug test?" I asked.

"You can always find somebody to piss for you. Even my wife found out what I was doing, and I still didn't care. Not even with two little girls at home."

Billy related all this as if he were merely telling another fire story, matter-of-fact and comic, the absurdity of someone else's tragic life. But it had the scripted feel of a twelve-step tale, as if the story had become more real than his memories. I could tell that he wanted to be viewed as a cautionary figure, and though I was fascinated, it all felt a little more earnest and preprogrammed than the normally twisted worldview I had come to appreciate from Billy.

"Anyhow, the department eventually caught up to me. They told me I'd be fired if I didn't go through rehab. My plan was to wait for the next good fire and step into the flames, just let myself fall through a roof or something. Then my family would be taken care of and I'd be a freaking hero, dying like a fireman. But the chief said he'd pay for rehab, so I gave it a shot—and don't ask me how, but here I am."

I didn't know what to say to him. This wasn't one of those situations where there was an established firehouse patter to fall back on. The typical "Nice going" compliment seemed a little pale in light of what Billy had just told me. I spun the fuel cap back on.

Billy filled the silence for me. "I tell you, it never goes away, though. I got shoulder surgery after a fire a few years ago. I'd been clean for seven years or some shit like that. And *bam*, within three days I was addicted to the Vicodin they were giving me for the pain. Three fucking days. I dumped the pills down the toilet and just lived with the pain until I healed up. I'd take anything to keep from getting addicted again. My body just can't handle it, so I don't give it a chance anymore."

Billy stopped talking and met my eyes. "You don't have any idea what I'm talking about, do you? The addiction, I mean. The *needing* it."

"No." I felt embarrassed, like a sheltered rich kid who doesn't know what real life is all about.

"You're lucky, man. I only have to look at a beer and I want to drink the whole fucking brewery." He patted his shirt pocket. "You want a smoke?"

"That's okay, Billy."

"What's the matter?" he said, grinning. "Don't you want to be just like me?" He shook a smoke out of the pack and lit it. Somebody had flipped all the cigarettes over when Billy wasn't looking, and now the filter was burning instead of the tobacco.

"Fucking bastards!" Billy spit the cigarette on the ground and went running into the firehouse, cursing and laughing, looking for somebody new to amuse himself with.

Police can't travel through the worst drug-infested neighborhoods with the same impunity that we can. The fire engine is a giant red force field of safety. There are neighborhoods in Oakland that I wouldn't dream of walking into in my street clothes. But when I'm wearing turnout pants and a T-shirt with OAKLAND FIRE emblazoned on the back, there's no place I feel threatened. Our dress uniforms are nearly identical to those of the police officers, and when I'm in my formal black button-down shirt, everyone mistakes me for a cop and I feel the hostility and mistrust. That sort of constant disapproval can't help but harden a police officer to the world, make him bitter and suspicious. But for us the opposite is true, and we can take comfort in the smiles of adults and the waves of kids when we pass by. People in Oakland know that our only interests are putting out fires and helping sick people, and we make sure to cultivate that trust. I've seen kids dealing drugs or cleaning their pistols on the corner without shame when we drive past. For everyone's sake it's better to turn a blind eye than to tell the police about the petty stuff. In some cities firefighters wear bulletproof vests; in Los Angeles they had to dodge bullets while fighting fire the night of the Rodney King verdict. But in Oakland things are different. Maybe I shouldn't feel so comfortable, but with the lack of any bad evidence to the contrary, it's good to believe that I'm welcome, that everyone knows I've only come to help.

Of course, as Billy proved, you don't have to live in a bad neighborhood to have bad problems. I don't know if drugs are the cause of tragedy or a symptom of it. But I do know that addictive substances underlie the majority of the calls we go on. If it weren't for alcohol, tobacco, and illegal drugs, I might be out of a job. Every day I have the potential for the wild excitement of a fire, but more often I'm called out to help someone who has fallen afoul of his or her particular vice. I get to use the Hurst tool to tear cars apart because most accidents are set in motion at the corner bar. Shootings and stabbings are the result of drug deals gone bad or boozy arguments that wouldn't seem worth killing someone over in the sober light of day. And I can't count the number of times I've had to ask someone to put out his cigarette so I can administer oxygen for an asthma attack.

We got called out to Oakland's fanciest downtown hotel one cold spring day. Usually we go there because the desk staff wants us to roust a drunk who's fallen asleep near the front door. But this time a security guard led us through the lobby and into the elevator. He turned his key next to the "penthouse" button, and when we got to the top, the doors opened to what I imagine Robert Downey Jr.'s life must look like. The luxury suite was a mess: Broken bottles of liquor lay in the sink, a bloody pillowcase was on the floor, and the whole place smelled of vomit and pizza. Two supermodel types with vacant expressions lounged on a couch, and I couldn't help but wonder what was going wrong in their lives that they had ended up like this, in Oakland of all places. The object of our attention was a man of about thirty-five, with very nice (but freshly soiled) slacks and enough oil in his hair to grease a fleet of fire trucks. He was sitting on the bathroom floor, leaned up against the toilet with his head rolled back. When I checked him, I saw that he was breathing maybe twice a minute and his pupils were pinpoint small—classic signs of heroin overdose. I pride myself on my diagnostic skills, but the needle still dangling out of his arm wasn't a bad clue either.

Heroin overdose is an easy one. I put a short tube in his mouth, just long enough to tickle the back of his throat and deliver some oxygen. One injection of Narcan to his shoulder, and then there was nothing to do but use the ventilator bag to breathe for him and wait. Narcan goes straight to the chemical receptors where heroin is received and blocks them out. True to form for resuscitated junkies, in about two minutes our playboy started waking up, gagging on the tube, and cussing us out for stealing his high. Another satisfied customer.

About an hour after that call, we were sent to a grimy, urine-smelling park three blocks from the hotel, a place used exclusively by dealers and homeless people. A young guy was lying on a park bench, and somebody told us that he'd been "stuck in the ass." It didn't look too bad. Just your average ass stabbing, I guess. I took some gauze and tape and started to apply pressure, but he swatted my hand away and whispered to a guy standing next to him. I tried again: same result, more whispering. I told the guy that I wasn't a cop and didn't care what he was doing, but that if I didn't stop the bleeding and start an IV soon, he'd really be in trouble. A little more whispering, and finally the truth emerged. It turned out that his butt was not just an injured body part but also a rich warehouse of tiny Ziploc bags full of crack. He wanted to make sure they were secure before he went to the hospital or to jail. I told him to do whatever he wanted, as long as I could get to work soon. So he lowered himself to the ground, knelt in front of the bench, and offered his bleeding rear up to his friend. I gave the friend a pair of gloves and, as if it were the most normal thing in the world, he went to work extricating the booty. Problem solved. The friend disappeared into the night, and the patient was compliant as could be. We did what we needed to, loaded him up onto the ambulance, and sent him on his merry way.

Oakland's just like any other big city, and like most small towns, too, I suspect. So many of our intractable problems lead right back to the stranglehold of drugs. We don't have gangs per se, no Crips and Bloods who battle for total domination. Instead we have neighborhood associations, small blocks of turf staked out by local dealers and their friends. When I was a kid, the high school—the public one, not my insulated haven in the hills—was forced to ban book bags. Too many kids used them to carry drugs and guns. The murder rate is cyclical, perhaps tied to the economy, perhaps not. After the market bubble burst, the homicide rate jumped 60 percent in one year; I don't think the murderers were worried about the price of their Microsoft shares. Newspaper stories play up the carnage of the drug trade, and the homeowners in the hills fret about how unsafe the city is becoming. But the people in good neighborhoods don't have anything more to worry about than they ever did. There's never an episode of violence that sweeps the city indiscriminately. The violence is always circumscribed, confined to specific locations where most middle-class people never go, places that I never knew existed before I got this job. The Ninety-eighth Avenue corridor. Thirty-fourth Street and San Pablo. Carney Park.

And sure it's racial. I'm not so self-righteously progressive as to pretend that I don't see color. I can't miss the fact that the African American community in Oakland has been visited by a disproportionate share of suffering. I see that, but I can't change it. I don't have the cure for drugs or violence. I know that the root cause is always poverty, but I don't have a single job that I can hand out to anyone in this community of mine. I do have a skill, however, and the will to use it. I am given the choice every single day, as to whether I'm going to treat all my patients with the same concern and professionalism or cherry-pick out the ones I like, the ones I'll give the best treatment to.

It's lofty to say that you'll treat every person the same. On the street it's hard to stick by your principles. I do the best I can, but I'm not as blind as I'd like to be. When it's three in the morning and you've already picked up the same guy three times that shift, not to mention every single day in the last month, it's hard to feel generous. When a patient is drunk and abusive, when he smells of vomit and curses you, it's hard to want to help him.

But nobody wants to be sick. The drinkers, the dopers, the gang-bangers—there's not one of them who doesn't wish that he could change his life to something better, something easier. They all wish that a fire engine wasn't the only thing that could hold out a thread of hope and safety. Nobody wants to be lonely and cold and afraid. Nobody wants to be the man that no one wants to help.

One paramedic I know had business cards printed that read, REMOVING THE SICK AND INJURED FROM THE PUBLIC EYE SINCE 1987. There's a feeling among some people in the department that we're just another link in a useless social-welfare machine, fighting an unwinnable battle against urban decay. But I don't feel like that. In an odd way, this never-ending struggle against poverty and despair is heartening to me. I am the closest thing to universal health care that may ever exist in this country. No matter your income, education, or tragic history, just call 911—and in three minutes four highly skilled firefighters in a gleaming, half-million-dollar engine will pull up at your door. A lot of people call us because they can't afford a cab to the hospital. A lot of people go to the hospital because they want a hot meal and some attention. Some people need that attention so they can survive life on the street for another hour, another day. It doesn't matter; we don't make judgments. They call, we respond.

Getting My Brown Belt

Whatever you do, don't stand too close to Captain Powell at a fire," Tom said, looking up at me from down on one knee, where he was putting some tire dressing on the rear dual wheels.

"Why's that?" I asked. Washing the rig was a good time to pick Tom's brain. I'd follow behind him with a soapy rag or a glob of wax and try to coax him into conversation while we went over the truck on a Sunday morning. He'd been telling me stories of how firefighters get hurt.

"He's a hell of a fireman, but he's cursed." He let the statement hang.

"It seems like he knows what he's doing," I said.

"Sure he does. But I can't think of anyone who gets hurt as much as he does. If you're on his tail, you're liable to get nailed right with him."

Billy had been listening to our conversation as he washed the engine in the next driveway over. He'd been lighting each cigarette off the previous one, and despite the bags under his eyes from six straight days in the firehouse, he still moved with his customarily jerky, frenetic mania. He butted into the conversation with a cough.

"See, Pops, that's the thing about you."

"What now, Billy?" Tom said with practiced exasperation.

"You're old, and you don't think straight." Billy turned toward me. "You've got to be smart about this stuff. What you want to do is stand as close to Captain Powell as possible. The thing is, usually only one person gets hurt at a time. And if somebody's going to fall off the roof or get hit by a beam, you've got to figure it's going to be Powell. If you stand next to him, you've got his bad luck to protect you. If I were you, I'd stick to him like stink on shit." Billy smacked his hands together and lit another cigarette.

Tom stood up slowly, one hand on his back, and looked at Billy from underneath his heavy eyebrows. He shook his head like an indulgent grandfather and turned his back on Billy, moving forward to the next set of wheels. Billy coughed and flicked a long stack of ashes from his cigarette into the bucket of water that Tom was using to wash the truck.

I was surprised to hear that Captain Powell was so accident-prone. To my eyes he'd always been the model of poise on the fireground. He was humble to the point of being unduly apologetic and usually started every phrase with a long "Weeeelllll," as if to say that he didn't want to be presumptuous by actually having an opinion that might be worth listening to. He was visibly uncomfortable in dealing with the personality clashes and personnel politics required of a captain in the modern fire department. In general, the guys were forgiving with his hesitating, unsure manner at roll call because of the remarkable transformation that occurred in him as soon as the bells tolled for a working fire.

On the fireground he just moved right. I never saw him fumbling with his gloves or struggling to find the best way to arrange a load in his arms. He always seemed to be carrying exactly the right tool, and he placed it perfectly the first time, every time. His truckman's belt and radio holster were sewn directly into his turnout coat, so that the myriad gadgets and appendages seemed like easy extensions of his body, as natural and quick as his darting eyes.

The coolness of his command had earned him the unambiguous respect of both firefighters and chief officers. I once heard him give a size-up of a fire over his radio as calmly as if he were ordering a pizza: "Oakland Fire, this is Truck One, we have a five-story residential, fully involved with flame. Multiple citizens in need of assistance [there were at least half a dozen people hanging out the windows, screaming for their lives], and we'll need additional resources. Please strike a second and third alarm. Thank you, dispatch." Please? Thank you? These words rarely even heard around a firehouse dinner table seemed so out of place in the impolite world he was describing over his radio. His entire demeanor conveyed calm, and I felt safe working alongside him, even if I didn't understand his plans until he explained them to me long after the fire had been knocked.

Powell's gift was his ability to visualize the totality of a fire scene, his knack for stripping away the inevitable chaos in order to reduce to the essentials what had to be done. It is almost impossible not to focus on the area with the biggest flame, but Captain Powell was able to think three or four steps ahead of a fire. Is that power line close to burning through? Is the gas station across the street at risk? Do the men on the roof have three escape routes or only two? I often saw chiefs turn to Powell for advice, and he would offer it without hesitation, then fade back to his crew without seeking praise or recognition.

"I can't imagine what he's still doing here," Tom said, scratching at a bit of road dirt with his blunt thumbnail. "If I got hurt as much as he does, I'd retire—or at least transfer up to the hills."

The image of Captain Powell as a bumbling oaf didn't fit with my mental image of him. "What does he do?" I asked. "What happens to him?"

"Well, let's see," Tom said. "He walked through a skylight and rode it down three stories to the ground. He buried a chain saw in his thigh. He tore his knee all to shit—"

"Don't forget about the finger!" Billy chimed in.

"Oh, right," Tom said. "He ripped all the skin off his finger when he caught his wedding ring on something stepping down from the truck."

"And the motorcycle thing, too," Billy said, like a precocious schoolkid bursting with all the answers.

"I guess he went for a dirt-bike ride with the guys," Tom said, starting to warm to the history of his captain. "He wiped out coming around a hill, but he was too embarrassed to tell anybody, so he just pushed his bike all the way back to the truck. By the time he got back, I guess he was pissing blood, but he didn't want to bother anybody, so he just lay down on the seat and passed out for a while until he felt better and called his wife on the cell phone. He was just going to take a couple of vacation days and not tell anybody, but somebody ran into his wife at the hospital and figured it all out. Fuck, he's tough as shit, but won't tell anybody anything."

"He won't give anything up," Billy said. "You can't *buy* a story out of that guy. But if it's happened to somebody, it's happened to him. He even *died* in a fire one time." Billy took a deep suck on his smoke and waited for me to ask.

"What do you mean?" I asked.

"I dunno. It was before I came in. You were there, right, Pops?"

Tom nodded and looked at his feet. "I wasn't there that day, but I came on duty the next morning." He seemed not to want to talk about it. I'd heard hushed rumors about the fire in question and Captain Powell's heroism. Now I hung impatiently on Tom's silence, hoping desperately that we wouldn't get a run that would interrupt him.

"Go on, Pops, tell him," Billy prodded. "You do remember *some* things, don't you, old man?"

Tom straightened up and flicked his rag a few times, sending droplets of water and wax out onto the pavement. I could see he was reluctant to tell the story, and he sized me up before he started. It seemed as if talking about the dark moments of the job took something out of him, and he was trying to decide whether I was somebody who might stick around for a while, somebody who was worth his energy. He started slowly, with Billy chiming in details that he'd heard over the years.

It was 1979, and a fire had started on a BART car, the subway that runs deep beneath the bay to connect Oakland to San Francisco. Evan Powell, a young firefighter at the time, and his lieutenant, Bill Elliot, had entered the tunnel at one of the stations and were making their way toward the train along with several other firemen. They were wearing half-hour bottles, standard at the time, and on the way in, they stayed low and breathed the smoky air rather than exhaust the supply on their backs. As they approached the train, the smoke grew thicker, and they masked up in near pitch-blackness.

The fire raged over the length of an entire subway car, but miraculously, the firefighters were able to extinguish the fire and evacuate all of the passengers onto a waiting rescue train. At the time, BART was a new system, less than ten years old, and it appeared that the first major incident was going to be a textbook success. But the transit authorities made a serious miscalculation. As the rescue train sped away toward daylight and salvation, it pulled with it the toxic smoke and gases from a thousand burning seat cushions. The cloud followed the pressure wave of the train and charged the bore with noxious smoke. Elliot, Powell, and the other rescuers were plunged into blackness. The few feet separating them became an unbridgeable chasm. In the darkness and chaos, communication was nearly impossible, and each man was forced to make for himself a critical choice: Oakland or San Francisco? A walk in the right direction would mean clean air and daylight, and the other would bring only thickening smoke.

There would not be enough air left in the bottles to walk through the haze of a wrong decision. This would not be a choice, but a guess. The system was behaving erratically, so there was no way to intuit the route to safety, no experience to fall back on. Alone with the moist rasping of his own breathing, Powell made a decision and walked toward Oakland. Walked toward what he knew best.

The smoke only grew denser, but with the decision made, there was no going back. It was not how a fireman imagines he will die in a fire. There was no rush of flames, no searing heat. Just a cool, smothering darkness and the feel of a metal walkway stretching endlessly beneath his feet. There was nothing to do but keep walking until he saw light or felt the sudden tug at his lungs of an empty respirator. Being a small man and in prime shape, he used his air slowly. He consciously slowed his breathing even further, trying to walk lightly and sip sparingly from his only connection to the world beyond the tunnel.

Still feeling his way along blindly, trying to convince himself that he'd made the right decision, Powell felt a soft mass on the ground. Lieutenant Elliot had made the same choice as Powell, but his bottle had run out, and he had collapsed. He had unscrewed his air hose from the tank and plunged it inside his coat, hoping for any bit of filter from the toxic smoke. It hadn't been enough; he had succumbed and now lay on the walkway gasping, barely able to move.

Powell's reaction was instantaneous and without consideration. He held his breath, pulled the mask from his face, and pressed it to Elliot's. The fallen lieutenant took a few deep breaths and passed it back to his young subordinate. Buddy breathing in this manner, they were able to stagger farther down the tracks, farther toward Oakland, deeper into the smoke. They both probably knew that they had come up empty and made the wrong decision. But there was no turning back, and they continued struggling forward until at last, under the strain of two sets of lungs, Powell's bottle gave out as well. At that moment several other firefighters who had entered the bore were just beginning a miles-long walk to San Francisco, oblivious to the struggle of the men who had entered the tube alongside them.

Elliot and Powell came to rest, ceased breathing while leaning against a door that led to a gallery between the eastbound and westbound tracks. Topside, a rescue team had been hastily organized and was now making its way through the gallery. On a hunch, one of the rescuers forced open a door to the track and reached in with a long pole, sweeping it back and forth, until it contacted soft flesh.

The two men—bodies really—were pulled out. Neither was breathing, and no pulses were detectable. The rescuers found themselves in the position that firefighters dread above all else, and frantically they began CPR on their own friends. The fallen firefighters—"victims" now, a word normally reserved for civilians—were transported to the hospital. One of them was revived and spent the night in a hyperbaric chamber, the same kind used for deep-sea divers, intended to replace the toxins in his lungs with clean, pure oxygen. The next morning, his system flushed clean and his heart restarted, he emerged from the chamber, groggy but breathing.

Bill Elliot died that day. His badge is on a plaque at admin. Evan Powell died, too, but he lived to tell about it. And under his tutelage, a new generation of firefighters learned to stay calm in the face of pressure, to plan two escape routes, to find a way to rely entirely on oneself and entirely on one's team.

"That's why you want to stay away from him," said Tom, wringing his towel through his hands.

"That's why you want to stay close to him," said Billy. He eyeballed the hole at the top of his empty cigarette pack before crumpling it up and throwing it onto the ground.

Tom picked up Billy's trash. "Some of the guys who were down there that day ended up with lung cancer. The rest transferred right on up to hill houses. I don't know what the hell Powell's still doing down here; he shouldn't even be alive, really. He's the unluckiest lucky bastard I ever met." Tom tossed me a rag. "Go dry the front. I hate it when there's streaks on the mirrors."

·   ·   ·

The rigs were starting to gleam. They stood in a line crossing the sidewalk—the truck, the engine, the rescue, and then the chief's car, parked off to the side. The lady in the brick building across the street was hovering in her window, scowling over the flowerpots. She always complained when we did our morning checks, but that's the price you pay for living across the street from a firehouse. First thing every day, we check the sirens and the horns, we rev the engines at high idle to make sure the pump and the aerial ladder still do what they're meant to. On Tuesdays and Saturdays, we kick on every power tool, every circular saw and blower and generator, to make sure they're fueled up and ready. It's loud, and it's annoying; I know I wouldn't want to live next door to a firehouse, though living inside one a few days a week isn't too bad. We made a truce with our neighbor: no chain saws before 9:00 A.M. on weekends. Unless we have to.

I'm not quite sure why we spend so much time washing the fire engine; everybody always asks. I've owned a red pickup truck since I was sixteen, and it's never had a wash. Except once when another firefighter wet it down behind my back in the firehouse parking lot because he couldn't stand to look at it anymore. I've always figured a car is just a way to get from place to place, and as long as it runs, I don't much care what it looks like. But washing the fire engine is the most archetypal "fireman" thing that we do. It's what everyone associates us with. That and grocery shopping. And carrying babies down ladders.

A fire engine is special. It's as much a symbol as it is a tool. Little kids don't stare and wave when I drive by in my own truck; cops don't give me the nod; people don't stop what they're doing and crane their necks. But even kids too young to speak cock their heads to the side and drool a bit more fervently when a fire engine drives by.

A fire engine has to look good. The people pay us to be ready for them, and it's our obligation to maintain a professional high profile. Washing and waxing keep us out in the community instead of inside watching TV or reading the sports page. People on the sidewalk stop for a minute and chat while we're rinsing the soap from our hands. These fire engines belong to the citizens, and the people like to have a look at what they've paid for. It's the same reason I leave the bay doors rolled open during the day. If people drive by and see us, ready to go to work for them, they feel safer. They know that their city is watching out for them. In many of Oakland's neighborhoods, the fire engine is worth more than any of the surrounding homes. I want the rig to look nice; Oakland deserves it.

For a while Oakland and a lot of other cities around the country experimented with fire engines that were yellow or neon green. A professor somewhere did a study showing that drivers saw those colors more easily. But in practice, yellow rigs were a disaster. Not only were they ugly, but people never identified them with the fire department. People figured that they were being followed by tow trucks or street sweepers and didn't bother to pull over. It might be easier to see yellow, but red gets more respect. The ugliness alone was reason enough to get rid of those yellow monsters. You shouldn't mess with a fire engine. Fire engines should be big and red and have long wailing sirens, not the new computerized chuckles and chirps. If possible, the cabs should be open or, better yet, the firefighters should be standing on the tailboard. The chrome and the brass should sparkle, and the paint should be buffed so clean it looks like it's still wet.

But in reality that sort of near-mythic cleanliness rarely happens. At three in the morning, you just want to make sure all the tools still work before you crawl back into bed. You want to leave the next shift to hose the ashfall off the fire engine's roof and wipe the muddy handprints from the compartment doors. Federal worker-safety law now says that all new rigs have to have enclosed cabs and that firefighters can't ride on the tailboard anymore. But it doesn't matter. It's still a fire engine, and it still looks good whether it's shiny clean or dirty as hell from some smoky beating it took along with its crew.

On a nice summer morning, you can't beat the feeling of washing a fire rig—truck or engine, it doesn't matter. You can't help thinking, *This is mine. I drive this. I save lives with this thing and with these guys here around me.* On the morning I heard Captain Powell's story, despite Tom's admonitions and doomsday tales, I'd been in long enough to have a little confidence, to say to myself, *I've seen fire. I know what I'm doing.* I'd gotten myself permanently assigned to Station One and was beginning to feel like a normal member of the crew. I was telling stories that began with, "I remember this one job . . ." and ended with, "Now, *that* was a fire."

We were in the soap-down stage, the rig all white and frothy, when Captain Powell's voice came over the speaker. "Basement fire, everybody goes." Chief Whitman was out first, running to his car. Seventy-three years old and looking good. The truck and engine left simultaneously. I love that: two rigs pulling out at the same time, like racehorses cutting around a corner neck and neck. Suds were flying off in the wind.

Rolling through the red light at Thirteenth and Franklin, sirens blaring, we passed under the shadow of the Tribune Building, the comic-book-skinny tower that defines the Oakland skyline. I'd gotten to a point where I wasn't scared or even apprehensive every time a potential fire came in over the speaker, and I could now appreciate the things around me instead of concentrating on getting my gear right and my mind arranged. I looked up at the Trib Building to assure myself that, as usual, each of the clocks on the different faces of the building read different times. In the breeze a green-and-white flag flapped from the pinnacle of the tower, showing a pale background of an oak tree superimposed with the lone word THERE. The sad truth is that Gertrude Stein had Oakland in mind when she said, "There is no there there." It's indicative of Oakland's middling status in the national consciousness that our greatest claim to literary immortality is a phrase condemning the city and, moreover, a phrase that no one even remembers was intended to apply to Oakland in the first place.

I can't help it, though; there is no time when I feel more like I've arrived at a perfect *there* than when I'm riding on a fire engine and there's
smoke in the air. That's the moment when it all comes together for me,
when everything makes sense and I wish that anyone who ever wondered why I wanted to be a firefighter could see me now. When the rig
is gleaming clean and you step off while snugging your helmet onto
your head as you kick the ax up smoothly and slip it down into its spot
on your hip like a gunslinger. It's all worth it then—the dead guy in the
hallway that morning, the petty hassles with admin, the pain in your
back that seems to be there every day now. Because you're a fireman,
the closest thing there is in this world to being a superhero. And when
it's going good, riding the rig is like flying. All the other drivers are
pulled off to the side in their cars, and the siren is screaming, and kids
are waving. And you cut through the streets like a dolphin through water, dodging, playing, slipping back and forth from one side of the road
to the other, snaking around concrete islands, slowing a little at the
lights, and then powering right through with a heavy foot on the horn
pedal. *That's* what being a fireman is all about.

The engine turned a corner hard, and the momentum smashed me
up against the side of the cab. Billy made a small clucking noise into his
headset. "On the left! God*damn*—I'd like to have me a piece of *that*."
A young woman in a low-cut gray business suit was stopped on the sidewalk watching us go by, her hands over her ears to block out the siren
noise. Firemen don't let trivial things like a lights-and-sirens emergency get in the way of the important things in life; a pretty girl is still
a pretty girl. "What about it, Harpo? Wouldn't kick her out of bed for
eatin' crackers, huh?"

"Oh, yeah, she's something else all right." I clapped my hands and
made some approving caveman noises into my mike. If only my mother
could see me now. Twenty-five years teaching women's studies at a
community college, and now her son is wolf-whistling at chicks from
the backseat of a fire engine. We made a right onto Franklin, swinging

wide around a driver who was too paralyzed by his fear of our horn to pull over to the side. All down the block, thin wisps of smoke leaked through cracks in the sidewalk.

Basement fires are tough fires. Heat rises, and you're starting at the top; it's like trying to fight a house fire by descending through the chimney. On a regular fire, you can always get lower, can always hug the ground a little tighter. But in a basement you need to either back out or else take a leap of faith, pierce through the heat at the top, and hope like hell there's something cooler underneath.

I was on the engine that day: fire attack. Engine Twelve was in first and already had a line on the ground. Tom and another guy went to work on the big metal sidewalk elevator, trying to find another way to let the smoke out. Ventilation is tough with a basement. The normal trick of opening a back window and putting a fan in the front door won't work if there aren't any windows or doors. A basement is a big stone oven. Tom worked with an ax like a young man, swinging big, fluid loops over his head and bringing the pick end down with a solid clang.

Captain Powell stood near the front door talking on the radio. I grabbed a second line off Engine Twelve and made a beeline for him. "Go on down," Powell said calmly. "I'll catch up with you in a minute." He always left the strap on his helmet unbuckled; the fit was so perfect that he could lean over upside down and it would never fall off.

Going down the stairs was hot but bearable. We were masked for the smoke, and I had my collar up and earflaps down. I'm glad that no-body films us through the smoke, because we must look like the Key-stone Kops, bumping into walls, getting tangled in our own feet. At the first bend in the stairs, there was the predictable pileup and calls to "lighten the line!" I knelt on top of the nozzle to keep anyone from stealing it away from me and started to pull slack from outside. That's the one job most often forgotten: Nobody wants to stand at a corner and feed hose around it. But you really need only one or two firefighters at the nozzle, and if the hose is jammed under a car wheel or snagged up around a banister, then the attack team will just tug until they tire themselves out and never get into a position to make the stop.

We kept going down. The only illumination came from the tiny flashlights on our coats, which just reflected the black smoke that was inking its way all around us. At the base of the stairs, I could feel the room opening up. I couldn't see a thing but imagined it to be like other downtown basements. They can be the size of a whole city block, often divided up by chain-link sections belonging to different tenants. There's usually a furnace room, a bunch of little storage nooks, some elevator hardware, and piles and piles of crap that nobody wants upstairs.

"Grab one of these guys and take your line around left." Captain Powell materialized out of the gloom, two or three anonymous masked firefighters by his side. "We'll go around right until somebody finds the fire or we meet in the middle. No problem." His words were muffled through the face piece, but he spoke slowly enough to be understood, and I knew what he wanted. I couldn't see any flame, but the smoke was getting thicker. I moved off, trailing one hand along the wall in order to keep my bearings. Whoever was at the corner of the stairs was doing a good job; the hose came after me as if it were being fed off a spool.

When the wall ended and made a left, I followed it around the corner. I didn't want to get into the middle of the room, where I'd have no reference points. The wall made another quick turn and then another. In the dark I'd searched the perimeter of a closet. The heat was building, but slowly. There was fire burning somewhere, and we weren't getting it. Nothing to be all that worried about yet; I wasn't too much hotter than I should have been from the exertion alone. I kept going along the wall. When things got in my way, I moved them if I could, or else I went over or around, always keeping a hand on the outside, always returning to the wall. I didn't like the thin sound of my own breathing amplified by the rasp of the respirator.

After one trip around something—maybe an electrical control box—I reached down for the hose I was still dragging behind me. My hand touched the hard round of the inch-and-a-half line, but I couldn't find the nozzle. I felt my way along the line, first one way and then the other. I realized for the first time that I didn't have anyone with me. Maybe he'd fallen off a few feet back, or maybe I'd never had anybody to begin with. In the pileup at the base of the stairs, I had assumed that somebody would peel off and come with me. I never bothered to check.

No problem. By feeling along the hose line, I would either come to the nozzle or end up back at the stairs. I had my bearings; I knew where I'd come from. I headed out.

The hose line hugged the base of a wall, and I followed on all fours, once again glad that no one could see me. I seemed to be following the hose longer than I had on the way in, but time and distance are often confusing in a fire. Now it was definitely getting hotter.

The hose I was following made a sharp turn: straight up the wall and out of reach. I definitely hadn't come in like that. I started breathing a little faster, then slowed it down to conserve air. I took off one glove and felt the hose. Instead of touching a rough cotton jacket, my hand was on warm steel. I'd been following a pipe.

Being lost in a fire is nothing like being lost in the woods. There is no almost lost, no almost found. Just lost. Completely and utterly lost. And getting lost in a burning building happens immediately. I've been lost in the forest, and it's a gradual process, a state of mind that you can allow yourself to slip into for a while. There's always a feeling of *I think I recognize that tree* or *I'm pretty sure I'm moving in the right general direction*. A smoky room, by contrast, is entirely featureless.

I could tell that nobody had found the seat of the flames yet, because the heat kept getting worse. My regulator said that I had a thousand pounds of air left, about a quarter of my tank, five or ten minutes if I could breathe slowly.

During a fire I've never once thought of dying. Afterward, when I look back at what I've done, I get scared. When I look at my life on a quiet day and make plans for my future, I think about how easy it is to find yourself lost in a fire. But I never think this at the time; there's always too much to do. The good thing about the fear of fire is that there's no moment of anticipation, no standing at the door of the airplane waiting to jump. It always starts out with nothing: You're reaching for another grilled cheese or watching the Raiders game or checking the oil on the rig. On the ride you're busy buckling your coat and flexing your fingers to work the old stiffness out of the gloves. Then there's hose to pull, ladders to be thrown. At a fire there's so much information to be processed, so many split-second miniplans to be formulated and rejected, that there's no time to be scared.

I started following the pipe back out the way I came.

By the time my low-air bell started ringing, a grating jangle that rattled my teeth, I'd followed one pipe to another and then to another. I'd taken my gloves off to get a better feel of things, but it wasn't helping. I figured I must be near a boiler room, pipes running out in every direction. Trying to stay low didn't help much either; pipes kept running away and out of my reach. Tom told me once that if I ever got lost, I should forget about finding the right way out and hack my way through a wall with my ax. Not such great advice for a concrete basement.

There was nothing to do but keep stumbling. Sitting in place was not a solution, and I told myself that I couldn't get any more lost than I already was. I kept following a pipe. It didn't lead to anything I wanted, but it was better than casting off alone into the center of the room. What I felt more than anything was stupid. Stupid for not making sure I had a partner, stupid for losing my hose line, stupid for breaking every rule about freelancing and fighting fire alone. Not to mention the fact that I was completely useless, no help at all to the fire attack that was still going on somewhere around me. The sounds of men working bounced all around the basement, so close I thought I could stretch out and touch somebody's leg. But there was never anyone there when I reached.

The smoke in a good working fire is viscous and greasy. It's on the dividing line between liquid and gas, so thick it feels like you can grab it with a fist. It leaves your skin oily and your face streaked with black plumes of sweat and mucus. Firefighters love that look; it means you worked hard. One day I want to take a mason jar down to a burning basement so I can bring up a little storm inside the glass and put it on my shelf to show people what being inside a fire is like.

I kept crawling, motivated more by mortification than fear; the fear was lurking at the edge of my mind, though, ready to swoop in at any time. My back hurt. My knees hurt. My head ached. I reached a hand behind me and grabbed the low-air alarm bell on my bottle. It vibrated up my arm in an unpleasant way, and I had to do a kind of hop-crawl on my one free hand, but it was worth it for the quiet. I needed to stop the ringing so I could start making plans for what to do when my bottle finally ran dry. I was screaming and cursing inside my mask, *pissed* that I'd managed to be so stupid. My yelling made a weak-sounding echo inside my face piece.

I didn't like how much hotter it was getting. I didn't have a radio with me, so if the chief had ordered everybody out of the building, I wouldn't have heard it. I kept seeing the word "flashover" in my mind, just the way it had been written on top of the handout in the academy. Flashover is the point at which a room gets hot enough so that all the contents (firefighters included) simultaneously ignite even without direct flame contact. Captain Gold had put it simply when we were in the tower: "Flashover is not a survivable event." With the heat mounting, the memory of his admonition to get out the nearest door when you see a flashover about to happen made me curse a bit louder and pound my hand against the unyielding wall.

Every time they find a dead fireman inside a building, his body tells you that he died working, fighting, stretching for the last inch. I wasn't thinking about dead firefighters at the time, though, and I'm sure nobody who dies ever is. The only possible thought to have is that you're in a tough spot, but with a lot of muscle and some luck you'll be just fine. You crawl till you can't.

"Lose a contact lens?" Tom stood above me, all muscle and hair and jowls. "Get off your knees, clown."

Just as I reached the edge of lost hope, there were Tom's legs in front of me, and when I looked up far enough, his grinning face looking down.

"Fire's out—you might want to take off that bottle and do a little something around here." Tom's mask was off, and he had an ax in one hand and a pike pole in the other, ready to go to work.

He was right. Sometime during my personal odyssey, the fire had been extinguished and I hadn't even noticed. It wasn't hot anymore, and when I ran a hand over the smear on my mask, I realized that it wasn't particularly smoky either. Somehow I'd circled back to the bottom of the stairwell and crawled through a whole room of men working. I was so relieved to have been found that I pulled off my mask and sat at Tom's feet, taking big lungfuls of the still-smoky air. A shaft of light filtered in from the door at street level.

"You busy right now, Zac?" Tom asked, shrugging his shoulders in good-natured mockery. "You want me to bring you some milk and cookies, or do you feel like you want to come give us a hand?"

Tom reached down and helped me up. I'd been found as quickly as I'd become lost. I resolved never again to let myself get separated from the team; I was going to stay as close to Captain Powell as possible. At that moment his aura of accident was vastly preferable to anything I might encounter on my own.

"Everything okay?" Tom looked at me with his head cocked to the side.

"Absolutely, Tom. Never better," I said, taking the ax he held out to me.

When I came up into the light, the street was just another fireground. There was hose on the sidewalk, overlapping lines of yellow and gray spaghetti crisscrossing from the engine and disappearing into the basement. Billy was laughing, kneeling down to change out a used air bottle. Captain Powell was over by the chief's car, rehashing the fire, drawing diagrams, and losing himself in the details, the tactics, his overwhelming love of the fight. Powell always looked taller to me on the fireground than he did in the firehouse; real work inflated him to his true size, changed him from the haggard administrator he always seemed to be as he fidgeted around the firehouse, pencil behind his ear and clipboard in hand.

Two guys were on top of the engine, laying the hose neatly back into its folds. Everyone else stood in the street, a line of ten guys all eager to get a hand on the hose to make it look like they were working, but mostly just happy to be standing around and chatting. I unbuttoned my coat and felt the chill through my soaked shirt. Somebody underneath the fire escape yelled, "Ladder! Ladder coming down! Clear behind!" By the tailboard somebody else shouted to the guys in the hose bed, "Coupling coming up! Take a short fold, last length." I was shaking a little, finally safe enough to be scared. Not to mention embarrassed and angry with myself. Mostly scared, though glad finally to be outside in the sunshine with a bunch of guys working away like it was just another fire.

Fighting fire is about small humiliations and large glories. Somebody found the fire that day. Somebody put it out. Somebody got a slap on the back. And when they looked in the mirror at the black streaks on their faces, they thought, *I made the stop today.* As for me, I spent the whole fire crawling around like an idiot without contributing a thing.

They both stick with you, the sting of embarrassment and the glow of accomplishment. I told Tom about it that night when I came into the kitchen and caught him with a giant spoon and a half gallon of ice cream.

"Yeah," he said, rolling the ice cream around in his mouth slowly, "that's what they call your 'brown-belt fire.' You think you know what you're doing, but you still get your ass kicked." He gestured at all the empty seats around the kitchen table. "I mean, you go three months without a fire, and then you get three in a row that you totally smoke, and all of a sudden you think you can do anything. Like that place there." Tom rested his spoon on the edge of the carton and pointed out the window to the church across the street.

Ever since I'd come to Station One, that church had been my waking nightmare, with its steeply slanted roof nearly perpendicular to the ground and the tiny rounded cupola on the front end. At night, lights from inside the nave would shine through the red stained-glass windows, making it look as though a fire was just starting to take hold inside. Without fail I'd catch my breath every time I glanced over and think we were about to have to go to work there. The window near my bunk on the firehouse's second floor was just about even with the peak of the church roof, and every night before going to sleep, I'd look out at the steep pitch and imagine trying to get a foothold while reaching out with a running chain saw in order to cut a vent hole. Mostly, though, I'd pull the curtain and be thankful that Tom was asleep in the bunk next to mine and Captain Powell was tucked into the officer's room downstairs.

No two days at the firehouse are the same. That's the allure and the terror of it. No matter how much you think you know, the circumstances are always more random than your knowledge is organized. Fire is chaos given form. Any plan you make will be undermined; no strategy you've used in the past can be used again. Maybe there's a bus stalled in the street on the way to the fire, and it gives the flames an extra twenty seconds to breach a wall. Maybe the window is a little higher than it should be, the walls a little thicker or less thick. Even identical buildings are put together and populated differently. Instinct is crucial. And knowledge, strength, experience, and guts.

"That place scares me," Tom said, upending the empty carton to lick out the last bit of ice cream. "I've worked here forever, and I still wouldn't want to have to fight a fire over there. And as soon as you think you know what's what, they come up with some new thing to worry about, like subways and hazardous materials and trench rescues and all that other new bullshit." Tom paused and looked out the window at the church. "People are always going to get killed on this job. That's just the way it is. You're right to be scared. I don't care what any-body tells you—your most important job is to go home in the morning."

"I guess," I said. "I hate being useless."

"Get used to it. It won't be the last time. There are no black belts in this business. It might help you feel better, though, if you washed this." He handed me his dirty spoon and walked off toward the dorm, looking older than I thought he should have, feet shuffling, shoulders strong but stooped just a bit.

Captain Powell came around the corner into the kitchen for his customary secret midnight snack and jumped a little at seeing me there, as if caught with his hand in the cookie jar. He looked small to me again, and I could more easily imagine him wearing an accountant's green eyeshade than the blackened red helmet he'd been wearing when I came up from the basement that afternoon.

"Nice work today," he said. Looking at his wrist, he tapped the face of his watch and smiled. "Weeeellll . . . nothing to do now but go to sleep and wait for our relief to show up. Nice going."

"Thanks," I answered, and then I headed for the stairs up to the bunk room, wondering how many shifts, how many fires, how many near misses it would take until I could honestly tell myself I was doing a good job.

Marrying a Fireman

Have you been driving the fire truck again? I hate it when you drive." Shona was sitting next to me in my truck, her shoes off, with one foot on the dashboard and the other one out the window, just the way my mom used to sit on our marathon cross-country summer trips when I was a kid. Except that Shona had her hands braced against the dashboard in anticipation of a crash.

"Stay in your lane! Oh, my God—you are the *worst* driver." She slugged my arm in half jest and cringed as I swung the truck wide around a corner and into oncoming traffic. "It's not like you've got a siren . . . or anywhere you need to get to. You're going to kill me." She shielded her eyes from the road and focused instead on the radio, searching for her favorite songs, sugary-sweet eighties pop candy that she thought she knew the words to.

"Sorry. It's just habit," I answered.

"And a bad one." She laughed.

Tom had been teaching me to drive the truck. "Just pretend that no-body's steering the tiller bucket behind you, and you won't get into trouble," he'd said. "Make all your turns wide, cut a big square, and take all the lanes you need. A lot of times I just drive right on top of the dotted yellow; they'll still try to pass you, but at least you'll have a little more room to play with."

Billy, as usual, had a different, more reckless take on the issue. "Whatever you do, never make eye contact with the other drivers," he said. "If you look them in the eye, they'll think you're sane, and you definitely don't want that. Make that eye contact and they'll think that you'll avoid them. But if you just stare straight ahead, you'll scare the crap out of them. They'll think you just don't give a fuck whether you get in an accident or not, and that's exactly what you want."

I hated driving the rig and avoided it whenever possible. I liked the freedom of sitting backward in the firefighters' jump seat, the extra minute to shake myself awake, buckle on my clothes, prepare myself. Driving also entails so much more than driving. On the engine the driver is responsible for the water supply and the pump that keeps the hoses hard. If the engineer can't adapt to changing circumstances and get water to his crew, the guys inside can find themselves in trouble quickly. The driver of the truck also has additional duties. He must decide where to spot the apparatus in order to throw the aerial ladder at the proper angle, without hitting any overhead obstructions. The truck driver is also the leader of the roof team, the officer's deputy on the scary end of the ladder, the one responsible for cutting a hole and keeping an eye on the safety of the topside team. Driving itself was plenty difficult for me without having to think of all the other attendant responsibilities.

"Well, I don't care what they're teaching you at work," Shona said. She smiled. "I just want you to—" Whitney Houston came on the radio and interrupted her train of thought. I loved how singing (badly) along with the radio took precedence over any other thing she might have on her mind.

She was right. There was no emergency. I'd taken the whole week off work, and we were driving up into the hills to practice for our wedding, scheduled for the next day. It was a typical Bay Area summer day—overcast, gray, and wind-whippingly cold. I'd been a firefighter for almost two years at that point, and Shona's boyfriend for just a few months longer. Out of all the years of my life, these were the two in which I'd found the roles that would come to define me: firefighter and lover of Shona.

Shona and I were married in a Unitarian church; in Berkeley that's what nominal Jews like me do when they marry nominal Christians like Shona. My meager understanding of Unitarianism squared pretty well with the sort of limited, progressive practice of Judaism that I'd grown up in: tolerance, love of learning, celebration of family, belief in "something," but only if it makes you feel better.

Normally I'm quick to avoid any genuine introspection as to how I live my life, but there's something about a wedding that demands a bit of navel gazing. Driving into the hills, I thought of how differently I'd turned out from anything I could have imagined. I was about to be married in a church in front of a pack of firemen, my colleagues. I don't know why I should have been so surprised; I'd never been expected to do anything in particular, never been leaned on to "make something of myself" or go into the family business. In fact, the main thing my parents worried about in terms of my future was that they would exert too much influence over it. My father is a doctor, the son of a doctor; my mother is a teacher, the daughter of two teachers. And while they both love the lives they've made for themselves, they were determined to let me make my own choices. They were always loving and enthusiastic, willing and eager to support me in my latest crazy whim of the moment. My mother was so aggressively nonjudgmental that when talking about my future, she'd say, "Now, when you grow up and get married—or find your life partner, I mean . . ." I think I broke her heart a little by not offering her any major personality crises she could indulge with magnanimity. Fortunately for her, I have a gay cousin on whom she can lavish her tolerance. Among acquaintances she likes to drop lines like, "I had coffee with my nephew and his boyfriend the other day," then pause for a moment so she can be shocked—*shocked*—when an eyebrow is raised. Watching her son get married in a church would have to suffice to fill her tolerance quota.

Besides, the Unitarian church had such a beautiful space—a fragrant, eucalyptus-lined patio overlooking the bay—that I decided not to let myself be bothered by a couple of New Testaments that I saw stacked on a table in a corner. I did, however, assure myself that there wasn't a crucifix in sight, and I tried as often as possible to get everyone to refer to it as the Unitarian "place."

"What are you thinking about?" Shona asked as we walked into the "place" for the rehearsal. She squeezed my hand and looked up at me romantically.

I pointed to a corner. "I was thinking that a fire could run right up that curtain over there, get into the rafters, run the cockloft, and there'd be fire all through here in no time."

"This is our wedding. Don't you think that's a little morbid?"

"Really?" The thought hadn't even occurred to me. When I walked into the room, my first instinct was to guess what length ladder I'd need to make the roof, how many feet of hose it would take to reach a fire in the rooms around back. I thought about the candles we were planning to put on the tables, the location of the exits, the yards of coiled extension cords that the DJ would need to run.

It was what I did with every building. Every restaurant, every friend's apartment, every grocery store had become a potential fire scene for me. Lately every time I'd seen a new-model car, my first thought was about how I'd dismantle it if it were crushed along the side of a freeway with someone trapped inside. It didn't feel morbid at all, though. This wasn't the off-duty cop's inability to see an honest man anywhere. Instead it was just a new way of thinking, a sign that fire fighting was starting to leak into my soul, becoming less of a job and more an actual part of me. I tried to think back to when I'd started to look out at the world through smoke-colored glasses, and I couldn't remember a time when I hadn't. I know that it must happen to everyone, that we all look at the world through the lens of our own expertise, but I'd never even seen it coming. I've never wanted to be someone whose job is his whole life, but the truth is that it's not so bad to constantly run through scenarios of how you can keep your loved ones from danger or daydream about the best way to save a life.

Two years before that moment, I didn't know a thing about fighting fire. And I didn't know Shona. I didn't know the difference between a truck and an engine, and I didn't know the way Shona would curl herself into a little ball in the morning as I left for work and chirp out to me, "Have a kick-ass day in the firehouse, little Zacky," then immediately fall back to sleep. I didn't know that I'd be skilled with a cardiac defibrillator and an ax, didn't know that I'd be blessed with the love of this tiny, quirky, forceful woman standing next to me and holding my hand in a church I couldn't bring myself to call a church.

"I love you so much," she said. "Promise me that you won't ever get killed in a fire."

"I love you, too, Shona."

"But *promise* me. Can't you promise me you won't?"

The first time I'd left for work, I could see that Shona was scared, and she'd looked at me like I was a soldier headed off to the front. For the first few months, if I didn't call her within half an hour of the end of my shift, she'd work herself into a panic, imagining all of the horrible things that might have befallen me during the night. But over time, with enough repetition of my safe return in the morning, she'd come to accept my safety as the default, rather than always waiting for a phone call from the hospital or, worse, a knock on the door from some chief she'd never met. I always made a point to tell her funny stories about the people I saw when I was working. There was no sense burdening her with the superheated rooms, the collapsed roofs, the dark hallways full of uncapped syringes and human despair. And while we both knew that the trash fires I told her about only scratched the surface of what my life was like, it always seemed better to leave the scary stuff unsaid, a workable truce in the neutral zone between comfort and reality.

But there it was, her request still hanging. "Promise me, Zac. Promise me you'll never go anywhere." And on the eve of my wedding day, I knew that there was only one answer for her impossible question. And I knew also that every other fireman's wife had asked for the same promise at some point and that every fireman had answered as I would, not just to say it but because he truly *believed* he was right. And so I said to her, "I promise you, Shona. I'll never leave you," because what else can a fireman possibly say to his wife?

The next day, the day of our wedding, was one of those uncharacteristic Bay Area summer days when it's actually hot and beautiful. The fog burned off, and even the smog swept away early in the morning, leaving the view from the "place" crystalline and sharp. The pine- and redwood-lined streets of Oakland and Berkeley gave way to the sparkle of the bay, the sway of the bridges, and the glimmer of sun off the windows and pillars of the skyscrapers in San Francisco. The wedding went the way weddings go, with speeches and crying and overeating and old friends who've flown across the country to see you and then barely get a second to wave at you from across the room.

After dinner I saw my grandfather standing outside, watching the sun go down, his hand resting on Captain Powell's shoulder as if they'd grown up together in New Jersey instead of having only just met that day. I'm sure my grandfather was asking the captain about his history, his family, how he'd come to be a fireman, and what he hoped for with his children. It was my grandfather's favorite thing to do, to find the very core of a person, to lose himself in the story of a stranger. Grandpa Charlie was the kind of person who felt comfortable in any group, the kind of guy who knew he belonged somewhere—anywhere, really—simply by virtue of his own good nature.

They made an odd pair standing together, but I know that neither of them felt it. My grandfather was a man who destroyed the stereotype of the bookish Jew. He was huge and strong willed and mischievous, a Paul Bunyan type who reveled in his physicality. All my photos of him are with a fishing pole, a lumberman's ax, or the clarinet he played in any band that would have him. Even as an old man who'd begun to shrink a bit, he projected confidence and strength. He looked like an old fireman. Standing next to him, it was Captain Powell who looked like the scholar. I could just hear Powell downplaying his own accomplishments, passing off credit and praise to his crew, trying to hide the fact that he was the acknowledged master-of-all-things-fire in the city of Oakland.

When I made the final decision to accept the job of firefighter, my biggest concern was how I would tell Grandpa Charlie what I'd chosen to do. My grandfather embodied the conflict inherent in socialist Jews of his era. He loved the blisters on his hands, delighted in the ideal of the burly Jewish laborer, yet in the end he needed those close to him to be intellectuals. He was a high-school principal at Newark, New Jersey's tough, urban Weequahic High, which produced such luminaries as Philip Roth, LeRoi Jones, and, of course, my mother. He was a history buff and a New Deal Democrat; in the years that he managed a Jewish community center, he always made a point to offer employment to the victims of Joe McCarthy's blacklists. He told me that he cried when Julius and Ethel Rosenberg were executed, and I think it was because he knew that it could have been him and his beloved Carol, my grandmother, caught up in the evidence-free hysteria of the times. He'd been so proud at my graduation from college, sitting in the front row and beaming, that he was willing to overlook the fact that my honors thesis had been about cows and that I'd never dated even one Jewish girl.

"You have to tell him," my mom had said gently, prodding, as I sat in front of the phone in her kitchen on the day before the academy started. The phone call to Grandpa Charlie weighed more heavily on my mind than any fears I might have had about spending the rest of my life running into burning buildings.

"Couldn't you just tell him the next time he calls you?"

"Don't worry," she said. "You know he loves you."

But it wasn't that. I was never worried that he wouldn't love me. I'd seen him adapt before. My cousin, who was out of the closet on the business pages of the *New York Times* before he came out to my grandfather, was something of a shock to Charlie. And yet Michael was the conventionally successful one, the publisher and businessman. Soon enough my grandfather couldn't help but be proud, and he looked forward eagerly to his monthly issue of my cousin's gay-lifestyle magazine.

I wished my grandmother could be there to help him understand. But she'd died when I was ten. She would have understood instinctively the things that drew me to the fire department. She used to like to say that she'd been in the third grade for thirty years. She'd watched Newark switch from Jewish to mixed-race to black but stay a tough ghetto the whole time, just in different shades. And through it all she'd kept teaching third grade, because she loved kids and because they needed her, every one. In her own way, she was a firefighter, too, always on call, always unquestioningly there when someone, anyone, needed help. She would have been right at home in Oakland.

Mostly, telling my grandfather was daunting because it made the change in me *real*. It even felt like coming out of the closet, revealing that my innermost desires didn't quite match those of anybody else in the history of our family. With that phone call it would be final, irrevocable. I would be a fireman.

My grandfather was quiet on the phone when I told him. He let me spill it all out in a rush, a hundred reasons that I'd massaged and rehearsed to appeal to his social sensibilities. He took a minute to answer. I could hear Benny Goodman on the turntable in the background and imagined him sitting with some impenetrable brick of a biography open in his lap.

"It's certainly a different kind of choice. But will they even hire a Jew?"

"This is California, Grandpa. I don't think they care."

"You do know you'll be the only one, though?"

"That's fine," I said. Surrounding myself with Jews had never been even a minor priority for me.

"I know it's a great job. But is it a great life?" There it was, the crux of his worries, and mine, too. "In five years, after you've seen every kind of fire, what then? You'll be bored? And what about a wife? You're not going to meet many nice girls in the firehouse." The image of his grandson the fireman had taken him completely by surprise, but I could tell he was warming to it. He'd be tapping his big fingers on his knee, unconsciously keeping time to music while his mind was somewhere else entirely. I could picture him smiling slightly, concerned but proud. He'd be imagining the romance of it all, the chance for a grandson of his to be a genuine American *hero*.

"Couldn't you maybe be a teacher, though? They're the real heroes," he asked.

"C'mon, Grandpa," I said. "How could there be anything better than a fireman? One minute everything's terrible, and then I show up and make it better. You can't beat it."

"Well," he said, "I know you're strong enough—you got that from me. Your grandma always said you were going to be a mensch. And you want to help people. How could I possibly want anything better for you?"

A flush of adrenaline mixed with giddiness filtered through me. Talking to my grandfather, I could feel my own conviction growing. It was starting to feel more and more as if becoming a firefighter was a social mission, not just something that fulfilled my selfish thrill seeker's urge. My grandfather's hesitant but ultimately admiring acceptance meant everything to me.

"And I know you'll be happy. You've always been good at that," my grandfather said, laughing. "Just make sure you find a girl that's smart enough for you. And Jewish wouldn't hurt either."

Shona wasn't Jewish, it was true. But she was smart and loyal and funny and irreverent, and Grandpa Charlie loved her. Besides, she looked beautiful in her dress, with flowers in her hair; my grandfather had always appreciated a pretty girl. I didn't put it together until that day, but the two of them had the same spirit, a constant open-armed embrace of the world. My grandfather's elitist streak had made him worried that a fireman wouldn't be "good enough" for the right kind of wife, but he'd met Shona and realized that no matter what he might have thought, I'd pulled it off and found an amazing woman. It's ironic that he would have thought that this job might be too earthy for my wife, the granddaughter of hardscrabble Alberta farmers and Yukon pioneers. Her grandparents had been worried, too, unnerved when she struck out for the States, making her way through good schools until she'd ended up as an attorney in the heart of the dot-com gold rush. Pushing papers around, wearing a suit, she was as far from her frozen Canadian roots as possible. Sometime during our separate childhoods, Shona and I had switched destinies. A kid from her family should have grown up to be a fireman, and a kid from mine should have grown up to be a lawyer. But following prescribed paths hadn't appealed to either of us, and we'd seen in each other a person looking to push outward and try the edges of new things.

And in a way I found that being a fireman allowed me easier entry into my wife's family. *What do you do?* they'd ask. *I'm a fireman*, plain and simple. No struggle to describe one of the amorphous professions that the people I went to college with fall into. I never had to say, "I write grant proposals for a nonprofit consulting group focusing on fluvial geomorphology." When I first met Shona's grandfather, Hector, a contractor who built the bridges, roads, and dams of the Yukon, he asked me, "So how's the fire business?" And it was great to be able to say, "It's good, Hec. Tough work, but we're getting it done."

My grandfather died just six weeks after Shona and I got married. But he danced at my wedding, swung his second wife around in a spirited Charleston, and spent the evening charming my colleagues and friends while beaming from ear to ear. He played a round of golf just days before he died and told my mom that he hadn't hit the ball so well in years; he'd just bought a new suit and sent it to the tailor to get it ready for the upcoming season of parties and concerts. I loved the way he lived his life, full of plans and good humor and affection for the people around him—friends, strangers, people he had yet to meet. Seeing Grandpa Charlie laughing with Captain Powell, I could tell that he understood my choice, that I hadn't fallen so far from the tree after all.

And when the lights went down and the music started, any questions I'd had about the mixture of the particles of my past were spun away. Antoine danced with Shona's old roommate. Tom tried to convince one of my college friends that he should give up archaeology for a career in the fire department. And when our friend Lou, a flamboyant gay man in a red plush jacket, went whirling across the floor with his boyfriend, all the guys from the firefighter table hooted and clapped along in time with their stutter steps. There was nothing one bit strange here; we were all just at a wedding, everybody dancing, everybody laughing, nobody worrying for a moment about how things were *supposed* to be.

· 12 ·

Things I Wish I Didn't Know

Most of Jack Alvarez's cooking was one-ingredient stuff. Meat. Rice. Beans. Nothing fancy, just solid firehouse food. I tried to be in the kitchen as much as possible whenever he cooked, hoping to pick up a few tips here and there. For a while I used to watch Billy, but all his dishes were too elaborate: "You just squeeze the tomatillos through the cheesecloth while you're waiting for the onions to crystallize . . . not too fast now." Better to stick with someone I could reasonably hope to emulate.

"Get me garlic!" Jack yelled. I cracked open a few heads and started peeling a couple dozen cloves for him. Jack never bothered to wash his hands before he cooked. He'd just put a fresh plug of tobacco in his mouth, wipe a greasy handprint across his sweatshirt, and start flipping knives and slamming pots around. Maybe that was why everything always tasted so good. "Onions!" he yelled again. It takes a lot of crying to cut enough onions for a whole crew, and I could barely see the knife through my tears.

Eating a meal at the firehouse takes about an hour and a half. Or, to be precise, eating takes about ten minutes, and "family hour"—which is code for sitting around and bullshitting—takes up the rest. We don't do a formal breakfast, just cereal and leftovers from the previous night, but that's good for half an hour or so. Cooking lunch and dinner for ten guys is no small endeavor, so the prep starts at least two hours before each mealtime. Throw in half an hour for wandering around the grocery store and an hour for washing dishes, and we are engaged in eating, getting ready to eat, or recovering from eating for a solid nine hours every day. Of course, meals are always interrupted by runs, so you have to plan dishes that won't be harmed by a two-hour hiatus in the middle of preparation. No fancy soufflés, nothing that needs constant stirring or delicate spicing, and definitely nothing that will collapse should a herd of firemen go clomping past the oven in heavy steel-toed boots. It also doesn't hurt to develop a taste for foods that are cold, soggy, or congealed. You're going to eat it anyway when you get back from the call, so you might as well learn to enjoy it, or at least pretend to. And for God's sake, don't go over budget. There's always a self-satisfied taxpayer in the grocery store who peers into our cart and asks, "So what are we buying you boys today?" If they were buying, we'd have lobster every night. Instead we each pitch in ten dollars a day, which covers lunch, dinner, and all the amenities that the city doesn't spring for, like telephones and the newspaper. The less we spend on food, the more we have left over for goodies like a new couch or maybe the fight on pay-per-view. Every now and then, somebody will get industrious and plant tomatoes and basil and onions and garlic in the patches of dirt alongside the firehouse. When it comes time to pick them, though, nobody is much willing to take the risk of eating something that absorbs its nourishment unfenced on the streets of Oakland. I'd rather not think about what Dominique and her buddies are doing in my vegetable patch.

But even within those constraints, there's no excuse for bad cooking. Nothing will get a new kid into hot water faster than cooking a crappy meal—or not enough of it. When I was sent to fill in at a new firehouse for a day, one of my classmates set me up by telling me that "they're all vegetarians over there." Needless to say, my first meal was not well received, and it took a few shifts for me to scramble my way back up to acceptability in the eyes of my carnivorous crew. It's fine to experiment with new dishes, but only if the experiment is a success. When I first started, I used to hope for a bell just as I was serving the meal, so that if it was lousy, I could try to convince the guys that it would have been great but for the interruption.

There are a few people, though, who somehow manage to go an entire career without ever learning to do anything in the kitchen. Captain Powell was one of those.

"No way. Fuck you, Julia Child. I'll do the cooking." Jack Alvarez was not the type to sit idly by as his day's meals were ruined. The roll-call sheet said Powell was up to cook, but Jack wasn't about to let him. Now, I'm a bad cook, no doubt about it. Certainly below par by firehouse standards. But even I have a few old reliables: a solid lasagna, a passable marinade for shish kebab, and a damned good chicken Caesar salad. Captain Powell, on the other hand, has nothing. After decades in the firehouse, he's one of the most respected fireground commanders, but everybody hides when he cooks. When he goes on vacation, we all cook in his place to move his name so far ahead on the cooking list that we'll be in the clear for a while after he comes back. His specialty is fridge casserole; he'll take every nasty leftover lurking in the back of the fridge and build up a steaming tar pit burned to the side of a Pyrex baking dish. One day he baked cookies in the morning and pulled the standard trick of hiding them so they'd still be around after dinner. Nobody bothered to look.

"I'll do the cooking," Alvarez said again, staring at Powell. "Don't come near the kitchen today, Cappy, you got that?"

"I really don't mind. It's my turn." Powell gave Alvarez his nervous little half smile, looking bewildered. Could he honestly not know how bad a cook he was?

"I don't care if *you* mind. *I'm* the one who's going to have to eat the slop." Alvarez had been working with us the day of Powell's worst gastronomical carnage. The new chief of the department had been coming around to every firehouse to have a meal as a way of getting to know the guys. We'd gotten busy during the day, things had slipped past us, and the captain had jumped right in where he was supposed to and actually cooked when it was his turn. Somehow, horribly, there it was: the chief sitting in front of a gray poached salmon with boiled celery as a topping. We dipped into the rice pot and carved out triangles like slices of cake. Alvarez had lived through that once and wouldn't let it happen again on a day he was working. Firemen have long memories, and they won't allow you to forget your mistakes easily. They pounce on any sign of frailty and exploit it endlessly. Tell somebody that you prefer to be called "Robert" and you can count on being called "Bob" for the rest of your career. One firefighter revealed that he didn't care for chicken, so of course the guys cooked bird every single day at the firehouse for the next three years. When he'd take a day off, the guys would post pictures of themselves eating swordfish fillets and barbecued ribs. Alvarez wanted to make sure that Powell's weakness was appropriately highlighted.

The bells came in, and Captain Powell tore the printout and read it over the loudspeaker. "Structure fire. Engine goes, truck stays." The fire was a few districts over, in the territory of some other truck. But every fire gets three engines, so it would be ours even though we'd have to travel a little.

"You're on the piss wagon, Harpo," Alvarez said. He was built for truck work; I'd never seen him on the engine. "Get out. I'll finish the onions." I wiped my teary eyes on my sleeve and made for the rig. One of the other firefighters was downtown at a class, so we were running short when the call came. Billy jumped up in the driver's seat and started leaning on the horn before we were even out of the barn door. He's a different driver when we're on the way to a fire—more fluid, a little reckless, better. The smoke was visible from the firehouse but back toward the hills a few streets, and we had to go way down a long boulevard before we could swing back around and up into the avenues. The fire was barely above the lip of a hill, so we stopped at the base to hook the hydrant, the flames just invisible over the crest of the asphalt. I dropped the hose strap over the top of the hydrant, then ran around to get into Billy's rearview mirror and shouted "Go!" as loud as I could. The three-inch supply hose pulled tight against the strap as the engine drove forward, laying line neatly into the street and out of sight up over the hill.

The hydrant was the childproof kind, designed to keep a kid with a pipe wrench from being able to open it to make a makeshift water park. It had a large round cap the size of a dinner plate, with angled grooves along the edges that will fit only our special kiddie spanners. These kiddie caps tend to bind, so along with the special spanner wrench, I'd taken a heavy mallet and a piece of pipe to use as a cheater bar. Sure enough, leaning my whole body into the wrench didn't do a thing. I gave the cap a few good whacks with the mallet, then slipped the cheater over the spanner's handle. The cap resisted again, then gave way. I swept the hydrant outlet with my finger, clearing any leaves or candy wrappers that could make their way into the pump, then threaded the hose.

At the bottom of the hill, I could smell the smoke and hear people screaming. Other sirens were getting louder from different directions, but for the moment I was alone. I didn't want to charge the hose with water until the engineer gave the signal; if I gave him pressure before he was hooked up, he'd never be able to make his connection, and we'd be without water. With no direct line of sight, I started up the hill. Halfway up I saw Billy come over the top and wave his hand above his head like a cowboy pantomiming the act of roping: the call for water.

I ran back down the hill to the hydrant, clunky and awkward in my boots and air bottle. My helmet bounced on my head and settled ingloriously, cocked forward and slumped into my eyes, as I coupled the hose to the hydrant. A tiny, ancient Chinese woman stood on the curb a few feet away from me, leaning heavily on two canes. She smiled and said something I couldn't understand, but I noticed the brand-new baseball cap on her head that read MILLION MAN MARCH.

I opened the spindle just a crack, slowly to avoid a water hammer at the far end. Water hissed, clamoring to be let up from the underground main and sent out to the fire. The first flow wriggled its way uphill, gradually pumping the flat hose up to a full, hard round. I finished opening the spindle, then collected my tools and ran up the hill. I'd heard stories about firefighters having their wrenches stolen or, worse, their water turned off if they left their tools by the hydrant. Hard to fathom.

A car screeched around the corner and started up the hill toward where all the action was. "Hey!" I yelled. "This road's blocked! What kind of stupid are you?" Even as I said it, I couldn't imagine where I would have come up with that particular phrase. I must have borrowed it from Billy. The vehemence in my voice was unlike me as well; in my civilian clothes, I'd never say a word to four kids out joyriding on a weekday morning.

"We gotta get through, man," the driver said dully, looking at me through half-lidded eyes.

"No you don't!" I yelled, even stronger this time. We were already running a man short, and I was pissed off to be spending my time jaw-jacking with these clowns while the house was still burning. Somebody in the backseat laughed, and the driver gunned the engine. The car bucked up and slammed down hard as it ran over the hose.

"Assholes!" I shouted after them, for lack of anything original to say. Upraised middle fingers came out of three of the car's windows. The tires straddled the hose line up the hill, then bumped again as they broke off the track near the top. I held my breath as they drove over, worried that the weight of the car would break the line and leave the pumper without a water supply.

At the engine Billy had his hands full threading up the connection. Captain Powell had pulled the attack line by himself. He was making his way down the long alley of a garden apartment. I followed behind, throwing the kinks out of the line. Smoke was pouring from the doors and windows of the last apartment on the end. The fire had already vented itself, and flames were coming from the roof, raining down little embers.

The hose came up short. We made the base of the stairs and no more.

"There's an old lady in there," said a woman cradling a baby in her arms and swiveling her hips back and forth to keep it asleep. We tugged at the hose, hoping it was just caught on a corner, but there wasn't any slack to be had. The captain kicked in the front door as I ran back to the rig. The truck crew was coming the other way down the alley, carrying ladders and saws.

"We're short," I said as we passed each other. "Report of a woman inside."

I shoulder-loaded fifty feet of hose from the extra hose bin on the back. In front of the apartment stairs, the captain was standing by the useless hose end, jumping and twitching in his anxiety to get his hands on the extension. Two truckies were on the roof enlarging the hole; two more were inside making a search.

In our rush to couple the hose, the captain and I were fighting each other, four frantic hands making a mess of a simple task. "Hey, cut it out," I said, and slapped his gloved hand. When the connection was made, we picked up a firefighter from another rig that had just arrived and took the water up the stairs. With the smoke already vented by the fire and the work of the truck crew, the fire was easy to see. From the front door, I looked across the living room to the kitchen, where something on the stove had caught fire. The flames had quickly spread to the drapes, the cabinets, the walls. I opened the metal bale on the nozzle and adjusted the flow, not quite a straight stream, not quite a wide fog. The flames disappeared immediately. It might seem obvious, but it's always surprising to see what a good job water does in putting out fire. As we were getting a handle on the fire, the search crew hustled past carrying the slack body of an unconscious woman.

It was an easy fire. Smoke had filled the whole house, and the truckies had found the woman cowering, slouched limp against the wall of a small bedroom. The flames never did make it that far, but the smoke had, leaving thick black stains from floor to ceiling on the walls. With the fire out we pulled back a bit to regroup, to let the fire investigator have his routine minute before we started taking the place apart. I set down the hose and followed Captain Powell out to the landing. We turned off our bottles to catch a breath of fresh air, then stood watching.

The ambulance crew was waiting with a gurney in the alleyway. The firemen dumped the lady onto the rolling bed and backed away, stiff leather gloves turning the job over to crisp white latex ones. Somebody started CPR. One of the medics pried her mouth open with the laryngoscope, cursed, yelled, "Get me a smaller tube, maybe a six—she's closing up fast!" He dropped the smaller tube into her and listened over both lungs to make sure he was in. They strapped her in and rolled her away, threw the gurney in the back of the ambulance and drove off.

It's amazing what fire can do to things. A row of records congeals into a solid block and then melts over the shelf. Everywhere plastic hardens, freezing individual items together. Keyboards have only one key, a mugful of pens becomes a solid, fist-size mass. Whatever else is left is half mangled and covered in soot, and during the drudgery of overhaul, I challenge myself to guess what something used to be. Sometimes the spines burn off all the books on a shelf, and you have to open each one to see what a person's tastes were. It's a strange kind of voyeurism, looking at lives laid bare by fire. I often stop to see what's tacked up on the fridge, what magazines are by the toilet. In my mind I construct the lives of the people who live there, especially if they've died.

It's probably a violation of privacy, this re-creation of a life in my mind, but I can't help myself. I'll pretend to knock a book over accidentally so I can page through it quickly without being noticed. I often wonder how my own home would look after a fire. Would any of the firefighters care about the pictures of my wife and me, or would they just throw the blackened images onto the debris pile in the street?

In the beginning of my career, I had thought that fire fighting was about a *thing*, about gaining mastery over an inanimate process. But the truth I know now is that my job is unambiguously about people, about the complexity of the human spirit. Fire fighting is not about the fire. It's about the save. It's about keeping the flames from licking too deeply into a vulnerable life and wreaking havoc there. Our talent is not for putting out fire but for building tenuous levees of safety against chaotic rivers of destruction. When we open a nozzle, when we cut a hole in the roof, our job isn't only to extinguish the flames and lift the heat. What we're really trying to do is make a person whole again.

Something had definitely changed for me. Fires weren't all crisis and panic anymore. This time, for example, just about everything possible had gone wrong: We were a man short, the hydrant cap stuck, the hose didn't reach. And yet at no time did I feel that I was operating on the brink of disaster or at the limits of my capabilities. Even when disaster did strike and a woman lay near death, I was satisfied that I had made a competent—even damn good—effort to save her. It was the same feeling that had taken root in me after a few years of guiding white-water trips. I didn't always have flawless runs in the rapids, but I lost the unpleasant anticipatory sensation that I was more likely to flip my boat than not.

It was too much to expect that every fire would play out like a text-book exercise. Problems would arise, equipment would fail, people would die. In the drill tower, Captain Gold used to say, "It's not about whether you fuck up or not—because everybody fucks up—it's about how well you recover."

The woman who'd been carried out of this particular apartment was sick, I thought, and a good grandmother. It was a nice enough place, simple and old but impeccably clean, even after the fire. The remnants of lace doilies were on top of every chair, their ends curled and brown from the heat. Little ceramic figurines of Jesus and angels perched on every surface. The hallway walls were hung with fake wood plaques of ocean and sky scenes overlaid with trite homilies and devotions. Everywhere in the spare apartment was the apparatus of illness: a raised toilet seat with metal rails, a large box of pill bottles, a wheelchair in the kitchen. She must have left her chair and crawled into the back room where the truckies found her. The walls were covered with pictures of children, and there were toys on low shelves. I took framed pictures off the walls and stacked crinkly, blackened photo albums in my arms. I carried them down and set them aside from the growing pile of burned debris. I wasn't sure if the lady would live or not, but if she did, I knew she'd want these more than anything else she owned. It's always an awful thing to see a place burned to worthlessness. It's the people who have next to nothing who are always hit the hardest.

Upstairs the fire investigator had gotten what he needed, and now work could begin in earnest. Doorframes and ceilings were coming down in a search for hidden fire. Blackened heaps of everything were going out the window so they couldn't smolder and reignite. There is nothing more embarrassing than a rekindle, so we try to be thorough the first time. Even though that often comes at the expense of a person's walls and belongings.

There were about seven of us cramped in her tiny living room, each of us holding an ax or a pike pole or a shovel. I was tearing at a window frame with the flat of my ax, working the blade behind the molding, trying to avoid the butt ends of somebody's scoop shovel and somebody else's pike. We were all working quietly, a little subdued under the circumstances, without the usual hue and roar of too many firefighters in a small room together. But it was sort of artificial. Somebody had died, or was at least near death, so decorum required that we be quiet. But in reality nobody really felt sad or lonesome because we were still too many firefighters in a small room, still a bunch of guys who would go home afterward, finish lunch, and tell each other dirty jokes.

Tom pulled a charred wooden plaque off the wall and threw it onto the canvas carryall we were using to collect waste for the trip outside. As soon as it hit the floor, the plaque started singing. Or, more correctly, the fish mounted on the plaque started singing. The fish—it looked like a catfish to me, but the writing on the plaque said BOBBY THE BIGMOUTH BASS—gave a twitch, bent over double, and belted out "Don't worry, be happy" in a hearty Jamaican singsong. It was eerie, that singing. So Billy smashed it with his shovel, and we kept on working.

"Taaaaaake me to the riiiiver ... and waaash me off." The fish started in again on a new tune. Without breaking stride, Billy hit it a second time, scooped it up, and launched it against the wall. For the next half hour as we gutted the apartment, we left the singing fish off to the side on an incinerated couch. Every few minutes it would break into song, and somebody would kick it until it stopped.

It's hard to stay serious with a singing plastic fish in the middle of the room. The charade of sentimentality never lasts for long. Bobby the Bigmouth Bass made sure of that.

By the time we left the apartment, we were all laughing and making jokes and posing with the fish, singing along in falsettos. It was hard to remember that at the same time somebody we didn't know was at the hospital with a tube down her throat and smoke in her lungs. Maybe that's the point, though; maybe we can count on ourselves to be respectful and compassionate for only so long. Too much more would make the job unbearable. Firefighters are weird that way. I always think I should feel guilty for enjoying myself in a dead person's house, but it's rare that I can muster any genuine guilt. It's like the officer told me on one of my early fires: "You didn't start this. You're only here to make things better." It's strange, but I hope somebody will manage to get a laugh out of my death. I hope I'll be able to. A few years ago, when my grandmother was dying, she was being fed from a bright yellow bag that led to a tube in her arm. She looked up at the IV pole and said to me, "You know, I don't really care for this flavor."

Outside in the street, I stamped the ash off of my feet and ran a dirty coat sleeve across my sweaty face. Everybody was covered in grime but nobody cared.

"What do you think about *this* one? Should we call the medics back?" Billy had snuck up behind me and now tapped me on the shoulder. He was holding a tiny dog, maybe a ten-pounder, by the back legs. It dangled like something from a Chinese shop window.

"You're a sick puppy, Billy," I said, immediately wishing I'd picked different words.

"If only I'd gone to medic school . . . if only I could *do* something." Billy shook his fist at the sky like a soul-searching soap-opera star. The dog continued to dangle from his hand. I tried hard not to laugh. I wondered if Billy hadn't seen the unconscious woman who'd been pulled out or if he just didn't care.

Tom came up behind us and slapped Billy on the back of the neck. "Put the fucking dog down, Billy."

"What's the matter, Pops? I was just trying to help," Billy said, feigning indignation.

"C'mon. Just put it the fuck *down*."

Billy, not usually one to let go of a good joke easily, couldn't look Tom in the eye. Singing along with the fish in the apartment of a dying woman was one thing, but taking humorous advantage of a dead dog was very clearly another. Everyone has to draw his own line somewhere. Tom had two small dogs at home that he was always complaining about. He'd named them Piece of Shit I and Piece of Shit II. Quite a sentimental guy. But this was one step too far.

Billy laid the dog against the fence and propped a board over its little dead body so nobody would have to look at it anymore.

Fire and death are indiscriminate, cutting through all the artificial distinctions of race and class. Everything burns, everybody dies.

I remember one day in particular, a brisk, beautiful December morning, the kind of day that smells like Christmas, even to a Jew like me. Parked at the end of a woodsy cul-de-sac in the hills sat a single red pickup truck. Dispatch had sent us for a "man down" in his car, probably homeless with nowhere to go or a junkie looking for a place to sleep after hitting his high. Fred Calvin, our engineer for the day, parked the rig a ways back and, accompanied by the officer, started toward the truck with his rolling, stiff gait from too many years of bad fires and good softball. Fred was the sort of quiet easygoing guy that the fire department is full of; there's at least one of him for every screaming, eye-popping fireman on the other end of the spectrum. I like working with guys like Fred: They make me think that you can do this job for your whole life and not be changed for the worse by it.

I opened the side compartment of the engine, placed the cardiac monitor on the ground, and was just reaching for the airway bag as Fred looked through the front window of the truck at the end of the street. From a distance I saw him recoil, as if struck by the air between his face and the windshield. The shock lasted only a second, though, and then he regrouped and peered in again more closely. He looked up the hill toward me and drew his finger across his throat, the symbol we make when a call is canceled for any reason. "You don't need that stuff," he said, nodding at the medical bags in my hands. He walked past me slowly, trudging up the hill, hands pressed to his thighs against the pull of gravity.

I passed Fred on the downslope and took up his position by the driver's-side window. In the seat slumped a man in a heavy overcoat, his head leaned back so that he was invisible through the rear window. There was a shotgun resting in his lap, and the ceiling of the truck was dotted with dried blood. His face looked intact but sunken, as if he'd had all the bones removed from his right jaw and cheek and he was just waiting for somebody to pull the skin tight again over a fresh structure. His fingers looked cold and thin, blackened at the tips with the lividity of a days-old death. There wasn't even any point in hooking him up to the defib. The dark blood pooled in his fingertips meant that he had already begun to decompose, that he met one of the four criteria for immediate determination of death in the field. I've seen them all now, more times than I can remember: decapitation, obviously protruding heart or brain, decomposition, total incineration. The signs of hopelessness, the four signals that it's time to quit trying. One of the things I wish I didn't know.

The neighbors who'd come out onto their porches with the arrival of lights and sirens confirmed the time frame. They'd seen the man off and on for months, clearly living in his truck at the end of the road. This time the car had been parked for a week, and nobody had thought to look inside. It was the holidays, after all, everyone cozied up inside with loved ones while an unknown man cradled a shotgun in his arms.

Fred Calvin was five shifts away from retirement that day. If he's lucky, he'll have a long retirement doing what he loves, fishing and following the high-school hoops teams, and he'll never see another dead body again. Thirty-two years in the department and who knows how many hundreds of bodies, and he still recoiled, backed away with something squirming deep in his gut at the sight of fresh death. Prior to hitting the streets, I'd had no illusions that the job would be pretty; I'd considered car wrecks and cardiac arrests, prepared myself even for the sight of murder. But for some reason suicide had never entered my imagination. Suicide, while not something I see every day, has now become just another part of my shift, a run to be recorded in the company logbook like any other. Now I know the sad numbers, I've seen the spike in self-killing around the holidays. Women are more likely to attempt suicide, but men are more likely to be successful. Women prefer pills, seeking peace and an escape from violence even at the end. Men seek finality, reverting to an innate trust in machines—guns and automobile tailpipes. I wish I didn't know these things, but now that I do, I can't make myself forget.

13

Fireworks

The air was already thick and greasy as I left for work at 6:00 A.M. on Independence Day. The city already felt jumpy and anxious, when it should have been asleep. On New Year's Eve and the Fourth of July, Oakland turns into even more of a war zone than it usually is. The first New Year's I worked, I lay in bed listening to the constant rat-a-tat of automatic-weapons fire, hoping we wouldn't get a call that would send us out into the battle zone. But at the same time, I had half hoped for a chance to be in the middle of it, to be a hero racing in to make a desperate save. Oakland on holidays is what I imagine Beirut must have been like during its wartime heyday—lawless and dangerous, with bullets flying through the air.

On this particular Fourth, I was not disappointed. The fireworks in the neighborhoods had already been going on for a week, juvenile explosion freaks tuning up for the big night. As a kid I had played with bottle rockets and jumping jacks down by the piers, but the things I was seeing in Oakland were on a level with the pyro professionals. By noon I could stand on the roof of the firehouse and see colorful bombs exploding hundreds of feet in the air, rockets of flame and light. This year the game seemed to be to try to shoot the fireworks out of the air before they had a chance to explode, so every telltale whistle and whine was followed by a burst from a machine gun. I didn't want to think about where all those bullets were landing.

That day when I came to work, I bypassed the kitchen table and the coffeepot, went straight to my chores, kept busy and out of sight until the alarms rang. Shona had just had a miscarriage the morning before, and we'd spent all day at the hospital waiting for an ultrasound to confirm what we both already knew. I didn't want to tell anyone at work, couldn't stand a round of jokes about how maybe I wasn't doing it right and they'd be happy to come on over to the house one morning and show me. I felt fragile and stretched thin, like a rubber band that would pop with even a gentle tug.

So I swept and mopped and scrubbed and shined, throwing myself into the housework. I was hoping for a busy day, the kind that makes you forget you have a life outside the firehouse. Some days I like to sit on the apparatus floor and read a book or daydream, but on others I just want to get pounded. On the Fourth of July, I was hoping we wouldn't see the inside of the firehouse all day.

I was happy to see that my old drill-tower classmate Antoine Ugade would be the officer for the day. With injuries and retirements, the lieutenants' rank was running a little thin, and the administration had been letting qualified firefighters "act officer" for a day here and there. Antoine was a natural as an officer; he'd even acted like one when we were both recruits enduring our way through the academy. I liked working with Antoine; like me, he was a political junkie, and I loved his spot-on send-ups of Republican hard-liners. We would have been friends had we met in any situation; without the fire department, though, I'm sure I would never have had occasion to work alongside this young black man with a background so different from my own. It was hard to brood when Antoine was around.

"So what are you making me for lunch?" Antoine asked, coming up behind me as I listlessly swept the back of the bunk room.

"Old family recipe," I said. "Fried chicken, collard greens, black-eyed peas. Just like my grandma used to make in the old country. Maybe some sweet-potato pie, too, if you treat me right."

"Sounds good," he laughed. "Then I'll make some matzo-ball soup and latkes for dinner." The alarm bell rang. "Let's go, Shlomo," Antoine called out over his shoulder.

"Right behind you, Jamal."

At the address we walked into a courtyard that felt like the ground floor of a prison, three stories of railings rising above us in a square. The landings were jammed by men with cigarettes and stained undershirts, women with their hair up in rollers. Everyone just lolling, watching the fire department. And everywhere children—children running wild, children watching us from corners and rooftops, children flopping happily through stinking puddles. I thought about my own apartment, not fancy but clean and quiet. Shona had woven sweet peas and jasmine vines through the railings on our balcony.

A red-eyed man led us to the back corner of the courtyard where the garbage was and told us he'd seen a woman put a newborn in the Dumpster, said she'd delivered it herself and he was sure of it. We started to dig for what I fervently hoped we would not find: a lifeless tiny blue body. The two Dumpsters were overflowing, heaps of garbage on the ground surrounding each one. In neighborhoods like this one, the garbagemen come every other week, and then only if they feel like it. The piles of mess were steaming in the heat.

At first we pulled at the trash frantically, hurling wet plastic bags and chicken bones into the courtyard in our haste. Soon, though, we had excavated the cans to their midpoints, like archaeologists searching for last Tuesday. Nothing yet. But all action in the courtyard had stopped. The kids laughed and pointed when I jumped inside the Dumpster and started hauling out great pitchforksful of garbage. I watched my feet. I watched the pitchfork. I didn't want to find a baby that way.

The garbage wasn't just *on* me, it was *in* me. Chicken fat up my nostrils, clotted hair on the backs of my hands. Spaghetti on my boots, wadded tissue working its way under my glovecuffs. And flies everywhere, clouding over my head, mixing with the hoots and hollers of the crowd. I felt like crying. The stench-laden steam was rising off the garbage and filling every pore. My body felt as if it were leaking garbage, pooling filth in the bottoms of my rubber boots. I wanted to curse and throw things at the spectators. I wanted to crawl into the closed cab of the engine and hunker down on the floor, out of sight of everyone. Instead I just wiped my head with my arm and kept working.

Eventually we toppled the Dumpsters. We put a brick under the wheels to keep them from rolling, then leaned our backs into them until they fell. We brought the trash out with rakes and spread it a layer thin down the sidewalk. Every piece of paper got flipped, every plastic bag torn open and searched. The man who'd led us there pointed at a pile of chicken entrails in a paper bag. "I think it was that." We opened the bag and showed him. "Oh, sorry," he said, and walked off. We collected the trash back into a pile and then back into the cans. No babies, just a circus show.

They clapped for us when we left. They *clapped* for us, laughing.

Back at the firehouse, I wished I could boil myself clean. Instead I took a cup full of ice and water and sat behind the fire engine on the darkened apparatus floor, back where all the weights and barbells were scattered. When I'd left for work, I'd kissed Shona good-bye the way I always do. Sometimes she rolls over and gives me a sleepy hug, but that morning the alarm hadn't even moved her. She'd looked small and vulnerable, and I'd wanted to talk to her, but I hadn't known what to say. So instead I'd closed the door quietly and gone to work. Once we were back at the station, I thought about calling her, but whenever I had the phone in my hand, I couldn't get the numbers out.

When the doorbell rang, everybody jumped up, thinking that it might be a front-door still, a caller who bypasses 911 and comes directly to the source for help. As soon as the heavy roll-up door raised high enough to show off three pairs of bicycle tires, though, everyone else in the crew groaned and headed back to wherever in the firehouse they had come from. This particular troupe of kids had been coming by almost every shift for nearly six months. Our teach-the-children community spirit had begun to wear thin.

"Hi, Jack," said Jeremiah, the nine-year-old ringleader of his bicycle gang. My eye was drawn, as always, to the teardrop of a black birthmark against the lighter brown of his cheek.

"How you doing, Jeremiah? I'm not Jack, I'm Zac."

"Well, you look like Jack to me. I can't ever tell you two apart." Jack Alvarez is Hispanic and bald. He has a mustache and big arms, and he spits tobacco into a soda can. I look nothing like Jack.

"Will you fix my bike now?" Jeremiah said.

I flipped the bike over so it rested on the handlebars and seat and started pedaling it with one hand, feeling the chain jump with each rotation. The other two kids rode their bikes in lazy circles around the steep concrete driveway. Jeremiah always arrived with a shifting cast of cousins and friends, uncles his own age. One school day he came alone, and I asked him why he wasn't in class. He told me he'd been suspended for "sassing." When I saw him a week later, I asked if the suspension was still on, and he told me that he'd been kicked out for good. I didn't think they could do that to nine-year-olds, just turn them out into the street like that. I remembered the false smile on his face, all bluster and bullshit when I'd given him a "stop, drop, and roll" pamphlet that he couldn't read.

But this was the Fourth of July, a good day to be riding free in the city. In the firehouse we'd been on a holiday schedule ourselves: the bare minimum of housework, no drills, a nice wasted hour in the arm-chairs, and an A's game on TV. I pulled the bicycle chain aside from the gears and worked the two frozen links back and forth in my hands, adding oil occasionally until they swung as freely as the others in the loop. I flicked open my knife and cleaned some grit out of the derailleur with the point. The knife was a near switchblade thing that I'd bought because all the other firemen had one. It clipped to the side of my pants pocket, and I felt like an impostor every time I opened it.

"Can you lower the seat, too?" The bike he was riding was built for an adult woman, and Jeremiah's short legs barely cleared the sloped crossbar. I lowered it as far as possible, but he was still half a foot away from being able to sit down.

"Where'd you get this bike?" I asked. "I haven't seen you on it before."

"I found it on a fence," he said, leaning against my shoulder as he watched me work.

"Are you sure it's yours?"

"My cousin gave it to me."

"I thought you found it on a fence."

"That's what I said, Jack."

He rolled his head around in a big loopy nod, showing off the smile that was the reason I kept fixing stolen bikes for him, a different one each week. I wondered if he stole them himself or if they were gifts, or maybe even a business run by an older relative.

For a second I had a flash of Supercitizen in me; I thought about tak-ing Jeremiah under my wing, having him come to the firehouse every shift so I could teach him to read. But somehow I couldn't ask, didn't know how to make the offer. I knew he'd just laugh and call to his friends, and they'd ride off together. I gave him a bike helmet once, but I never saw him wear it. The day before, after the miscarriage, Shona and I had stood on our balcony watching kids playing in the park across the street. "Looks like we're going to have to torch the jungle gym so we don't have to look at that anymore," I'd said, trying a lame joke. Watching Jeremiah, I wished I could be at home.

When the alarm bell rang, I shooed Jeremiah and his friends away from the firehouse. He'd snuck in before as we were leaving, rolled under the doors as they were closing, so now we stayed parked on the apron until the door touched down on cement.

After lunch—BLTs ("Kosher for you, Antoine") with bacon gone limp and bread gone soggy from three interruptions—I was standing in the parking lot behind the firehouse looking at the teenagers shooting dice in front of the "Drugstore," a multiunit apartment building swarming with drug dealers and their customers. The streets on either side of the firehouse are long straightaways that people love to cruise in their cars with the bass pumped up so high that trees over the firehouse shiver and drop their branches on our roof. One time an overenthusiastic (or maybe just underskilled) driver lost control while spinning his wheels. He was trying to turn a rubber-burning doughnut but instead ended up with his car jumping the fence into our parking lot and landing roof to roof on top of the car of a firehouse visitor.

But on this particular Fourth of July, the streets were quiet. The sidewalks were busy, though; it was too hot to be in a car, and everyone had spilled out onto the stoops. Looking back down after a particularly good firecracker, I saw that the crowd of dice players was suddenly gone, disintegrated into the asphalt. Moments later a police car pulled up and disgorged a squat, middle-aged cop, who hitched up his utility belt with an audible grunt and disappeared up the steps. Thirty seconds later a man, maybe twenty years old, shirtless and slim, appeared on the stairs and flew down the street, his unlaced high-top sneakers barely touching the ground. The cop followed, flatfooted in boots and belly, his tools jangling loosely on his belt. I called to the other guys to watch the chase, and a few of us stood there, mirrored by the newly re-formed group of dice players, all of us craning our necks to watch down the street. Minutes passed, and the police officer appeared in the distance, walking slowly back toward us, his hand on the butt of his gun as if to prove that he was still the man with the power. "Tough luck!" I yelled across to him. He looked up but didn't acknowledge me. Jeers and taunts fell down on him from the windows of the building where he had just flushed his quarry. The city was buzzing. It didn't seem like a good day to be a cop.

By 6:00 P.M. it was clear that the city wasn't going to let us get any rest, and I was glad for that. It was better to be up and moving and working with the guys than trying to fall asleep in my bunk. I looked at the dispatch screen and saw that eighteen of Oakland's twenty-six engines were out on runs. Fireworks were landing on rooftops, setting ablaze gutters choked with dead leaves. Fights were breaking out at block parties, and the earth beneath the fire station shook with the reverberation of heavy explosives and the stereos of tricked-out cars cruising by. It had been the kind of busy I'd been hoping for. I'd barely had time to breathe, and Antoine had sought me out every time I'd tried to slink off to a corner. My dark mood of the morning had melted somewhat, and I felt vital and alive, as if my firehouse were the whole world. Some nights the city is simply electric, and we all gathered in the watch room to listen to the radio and wonder where the next fire would hit off.

The bell finally rang, and for the rest of the night we never lasted more than fifteen minutes at the firehouse before getting another run. Every time we finished a call and headed back toward the station, dispatch would call us on the air and turn us around.

A call came in for a gunshot wound about twenty blocks away. It was well out of our district, but since the two closer engines were already committed, it would be our run. Looking down side streets as we drove, I could see a haze of smoke, every alley looking as if there were a structure fire at the end of it. Little kids, probably not even old enough to tie their shoes, were lighting firecrackers one after another. The wind carried the whiff of sulfur and disorder.

Rolling to the gunshot wound, I reviewed the standard procedures in my head. A trauma call is simple but hectic. "Load and go" is the mantra, with the understanding that what a trauma patient needs is a surgeon, not a paramedic. My routine is always the same: Secure the airway, immobilize the spine, stop the bleeding. Start two large-bore IVs en route, run fluids wide open to bring back a blood pressure. For a gunshot wound or a car accident, I want the patient packaged and rolling to Highland Hospital's trauma center no longer than ten minutes after I pull up on scene.

Dispatch radioed to tell us that the scene was unsecure. If the dispatcher hears anything about a knife or a gun, she'll tell us to stage the fire engine a few blocks away until the police clear the scene. We hate staging. It's not in our nature to wait while somebody bleeds out around the corner. At a fire we wouldn't consider waiting until the scene is "secure." Crazy risk, that's what we're best at. It's all a matter of degrees. Prior to becoming a firefighter, I would have considered most of Oakland to be an unsecure scene; now I chafe when I have to spend ninety seconds standing back from a shooting.

We parked three blocks away as dozens of police cars flew past. Eventually the cops radioed to us to come forward, and we rolled in. On the ground in front of the engine, a man was sitting holding both hands to his knee. The shooting had taken place in the middle of a raucous block party, and the police formed a ragged ring around the victim, while five or six hundred drunken onlookers shouted and pressed inward. I gloved up, grabbed my gear, and knelt beside the victim, a

shirtless young man who was crying and screaming. As I reached behind me for my trauma shears to cut his pant leg, what I saw across the
street made me pull up short and forget what I'd started to do.

Ten feet to my left was a scene of utter carnage, more brutal than
anything I'd ever seen. A car was stopped in the middle of the street, its
sides riddled with bullet holes. In the front seat were two men, slumped
over toward the center, almost cradling each other.

"I'll be right back," I said.

The guy with the knee looked up at me as if he knew I wouldn't.
"Don't leave me," he said, pleading. Antoine met me at the car.

The men in the front seat were dead. They could not have been
deader. The driver was draped over the gearshift, his face shattered so
that I knew it was a face only from where it sat on his body. I looked for
entrance and exit wounds, but it was impossible to find order. The bullets had entered his body crisply, with a minimum of damage. Then, as
they are designed to do, the steel slugs had sought his bones and shattered them, turning his own body against him as shards of his ribs, his
kneecaps, his skull exploded and became secondary projectiles, leaving
gaping craters in his flesh. The wound that held my attention was the
least severe, one that had certainly not killed him. His left arm, still
gripping the steering wheel, was torn open from wrist to midbiceps, a
canyon of deep red that traced the layers of skin, muscle, tendon, and
bone. The wound was filled with broken diamonds of windshield glass,
packed as neatly as a fish freshly filleted and placed on ice. His anatomy
laid bare had a weird beauty that I tried hard not to appreciate. My eyes
kept returning to the deep red and flashing crystal of his arm as I
reached in below where his jaw had once been and pressed my fingers
in search of a carotid pulse I didn't expect to find. His skin yielded easily under my touch, and his muscles were just below the surface, young
and still taut.

"Got anything?" Antoine asked.

I shook my head.

"Check the other one. You never know." We both knew.

I walked around to the other side of the car, barely noticing the mounting anger of the crowd. Just minutes before, these kids had been cruising the block party, the smells of barbecue and weed in the air. Now this tall, slim boy lay back against the seat, his hand resting on his buddy's thigh, three holes in his chest. He had delicate, almost feminine features, his chin holding a few wisps, a boy trying to grow a man's beard. I held my hand to his neck, dug deep to find a pulse. I reached over him back to the driver, crazily thinking that I might have missed something, might have failed to catch one final, desperate beat.

"Nothing?" Antoine asked again. It wasn't really a question, so I didn't bother to answer; with the obvious damage, checking pulses was a formality. "Go get the other guy, and let's get out of here." Antoine turned away from me and said something into his radio, looking comfortable in the role of officer for a day.

Crowd noise filled my head for the first time. It sounded angry. People were shouting, "Why are you just standing there?" and "Aren't you going to put those people in an ambulance?" and "What the hell do we pay you for—help those guys!" There was nothing I could do for the two in the car. A man ran toward me, his arms windmilling as if he were trying to swim through the crowd. Immediately half a dozen cops were on him, wrestling him to the ground. It was all the crowd needed to erupt. The scene shattered. Any semblance of safety disintegrated. The gawkers who had once been partiers now became rioters. Scuffles broke out everywhere around us, and what had once seemed like a surplus of cops now looked like a dangerously small group. All around us the police were engaged in running battles with the crowd. Somebody lit off a firecracker, the pop sounding just close enough to gunfire to send the free hands of all the officers reaching for their guns.

"We're gone," Antoine said.

"What about the guy?" I was still trying to get my bearings, but Antoine was steps ahead of me.

"Put him in the rig, and let's go. We're not waiting for the ambulance."

Antoine called on the radio, yelling over the crowd, "We've got two DOAs and a critical trauma! This is a riot, and we're leaving the scene!"

"Hey. Hey! Come get *me!*" yelled the one who was still alive.

"I'm coming. Be right there," I said. I was scared but still self-conscious enough to feel ridiculous as I ran over to him in a combat half crouch that I recognized only from old war movies.

"We have to get you out of here. We can't wait for the ambulance," I said.

"Well, let's fucking *go*, then." He was shaking with the chills and trying to stifle sobs. His wild eyes kept scanning the crowd for whoever might want to finish the job. The pool of blood beneath him was spreading, and his knee was distorted grotesquely, bits of bone protruding at odd angles just beneath the skin, threatening to pierce their way through.

"Gimme a hand, Antoine. You've got the top." Antoine reached his arms around the patient and got ready to lift.

"You!" I said, yelling at a cop. "You come over here and lift his waist and that good leg." The crowd was still yelling, and the cop seemed paralyzed with his hand on his gun.

I held the bad leg in my hands, feeling it flex inappropriately.

"Stop stop *stop!* That hurts . . . you can't move me like that. Stop!"

"Don't stop, Antoine," I said. "Let's just get him up there."

"My leg! My leg! You're breaking it worse!"

"You're going to be fine. Try not to move." I was hoping he'd pass out.

"You're fucking killing me! Put me down *now!*"

"You're doing great. We're almost there," I said. Antoine backed up the stairs into the fire engine, hauling the screaming patient. We hoisted him up and deposited him on the floor amid our smoky turnout

coats and blackened helmets. He panted, trying to speak. I was dripping sweat.

"All right, we've got you now. No problem," I said. "We'll get you
some morphine here in a second."

Standing over him, I looked out through the front window of the
rig. Energy was still moving through the crowd like a wave. Pockets of
people were trying to break through police lines. Antoine jumped up in
the front seat, and before the door was shut, he was shouting at the engineer, "We're all on! Get the fuck out of here!"

It was a full-blown riot now. The cops were on their own. Antoine
turned to me. "Just like a block party in the old country, huh?"

I hadn't as yet given my one live patient so much as a cursory ex-
amination—it had probably been only five minutes since I first made
contact with him, but I felt as if I'd been in one side of a war and then
out through the back end. He was sweating profusely and speaking in-
coherently, his head drooping occasionally onto his chest. I didn't
know if it was the blood loss or the pain that was pulling him toward
unconsciousness. I couldn't feel any pulses at his wrists; his body had
shunted the circulation to his core organs. The pulse in his neck was
thready and fast, as his heart tried to compensate for the loss of blood
by pushing what little remained quickly through the arteries. I stripped
him naked with my shears, ran my hands down the length of his body
to search for any bullet holes I might have missed. I listened to his
lungs to make sure they were both still inflated.

After a few blocks, the driver stopped. The streets simmered with an
unnatural calm, and we could hear the roar of the crowd like the rumble of a distant wildfire. The ambulance pulled up alongside us. We
shoved a backboard under the patient, strapped him down tightly, and
laid him on the gurney. I gave a sharp knock on the back of the bus,
and the ambulance drove off to the trauma center. Ten minutes, two
lives, and one riot after first patient contact.

. . .

When we got back to the firehouse, I was tingling with adrenaline. This was exactly the Oakland that my parents had always warned me about. Antoine was talking to the battalion chief on the phone, telling her not to let anybody else respond to that neighborhood until things settled down. I reached for the private line to call home. The miscarriage the day before had been Shona's third. The one that meant that it wasn't coincidence, wasn't just bad luck; the one that meant the first inklings of fading hope. Shona's tougher than I am, less likely to throw up her hands and disappear when the going gets rough. When I asked her what we were going to do with ourselves, how much more of this we could take, she said, "We'll just do what we can. I can go twenty." She'd been crying when she fell asleep, though, and I hadn't talked to her all day. Now it was late again; she'd be back asleep.

I called my mom instead; I knew she'd be up, an insomniac like me. I told her about the shooting, the fireworks, the frenzy. I knew she could hear the gunfire crackling in the sky like electricity in a summer storm, and I hoped she would think it was only firecrackers. "It's great," I said. "It's been a great night."

"That's not what I want to hear," she laughed. "You're supposed to tell me that you spent the day barbecuing and teaching kids about Dalmatians."

"I'm sorry. I didn't want to worry you. It's just that it's such a *charge* out here."

"All right, kiddo. Make sure you guys take care of each other. Have fun? Is that what I'm supposed to say?"

It made me smile to hear her trying to be so positive in the face of what must have scared her so much. In return for her years of open-minded support, the choices I made for myself always seemed to require that she abandon the natural maternal urge to protect her young. My mom is every inch the Jewish mother, but she recognizes it in herself and works tirelessly to disguise her inner worrier. Wherever I am in the world, she mails me hamantaschen cookies on Purim and invites me home for matzo balls on Passover. When I tell her I'm being trained for confined-space rescue or nerve-gas bioterrorism, she bites her lip and smiles weakly, says, "That's wonderful, honey," and slides some fresh banana bread in my direction.

After I hung up, I looked at the fire engine sitting darkened on the apparatus floor. The air inside the firehouse was close and stagnant. The only sound came from the overloaded washing machine spinning itself unevenly with a cargo of towels and dishrags. There's nothing emptier than a fire engine with nobody in it.

When I saw that Antoine was done writing in the logbook, I sat down next to him in the watch room. He'd filled out the casualty reports for the two men in the car and was just straightening the desk. He'd changed into shorts and flip-flops and made himself a glass of iced tea. There was a nice breeze through the propped windows, and the neighborhood smelled like barbecue, smelled just the way the Fourth of July should.

"Antoine, if I ever—" I started.

"Don't," he cut me off sharply.

"No, I'm serious. If I ever die on the job . . ."

He turned his head to look me straight in the eye, almost angry now. "I don't want to be having this conversation with you, Zac."

"Trust me, Antoine, I'm not planning on it. But just let me say one thing. If I should ever . . . I mean, if it should ever happen to me . . . You'll look in on Shona every now and then, won't you?"

Antoine stood up quickly, almost upsetting his chair. "What do you mean, look in on her? You think anything I have to say could possibly make it all right with her?" He sat down again and drummed his fingers on his thighs. "Zac, you know you don't even have to ask. All I'm saying is that you'd better not make me have to do this for you."

"There's just one other thing, Antoine."

He rolled his eyes. "What is it?"

"If I die, I want you to promise me that there won't be any Jesus at the funeral."

Antoine's face softened for the first time and he laughed a little. "You got it," he said.

"I mean it. I want you to promise you'll tell the chief or the chaplain or whoever it is that I'm Jewish and I don't want any Scripture or hymns or any damn *Jesus*."

"So you want us all gathered in here listening to a rabbi?"

"I guess. Or maybe a justice of the peace. Do they do funerals?"

"Fuck if I know," Antoine said. "We'll figure it out if it happens."

"Thanks, Antoine—I'm counting on you."

"You got it, my Jewish brother. I'll make everyone wear beanies to your funeral." He offered his hand and shook mine firmly, encircling my wrist with his left hand.

Later that night, after six or eight more runs—wood-shingled roofs set aflame by firecrackers, an overabundance of booze and drugs for everyone on the streets—we were called back to the scene of the shooting. A lone police technician was pacing off distances, searching the sidewalks for shell casings. I wondered whether she felt scared, working alone at midnight in a spot where only hours before two young men—more like boys, really—had been shot to death. I felt hollow just looking at her. Had I been in her spot, I would have called the fire department back just to keep me company; anything would have been better than the artificial calm tenuously imposed by yellow caution tape flapping in the wind.

"Could you guys wash that down for me?" Barely looking up from her clipboard, she gestured to a dark pool of blood in the middle of the street where the car had been, where an arm had hung down, where an artery had come unseamed. The street was silent, and the car was gone, but the smell of firecrackers (or gunpowder, who can tell?) still hung heavily. I pulled out the redline, an easily reloadable, small-diameter

hose we use for minor jobs like Dumpster fires and smoldering mattresses. I turned the nozzle all the way to the right to make a tight
stream, then swept the water from side to side as I walked slowly forward, following the thickened shreds of blood until they hit the gutter
and floated downstream. The stencil on the curb by the storm drain
had a picture of a fish and the words NO DUMPING—DRAINS TO BAY.

We stayed a minute after the wash-down was done. The tech
showed us a few seconds of the crime scene on a digital video camera.
The film shot walked around the car deliberately, steady where I had
been breathing hard, at a professional remove from any hope of making
a save. She told us that the guy with the shattered leg had been a visitor from Chicago, that his friends had picked him up at the airport just
half an hour before he'd gotten shot. "Welcome to Oakland," said
the tech.

"Let's go," Antoine said. "We don't need to be here."

"You guys go on back to sleep now," she joked. "Quit distracting
me—let the workers work." Just looking at the rig made me feel good;
I wanted to be away from that street corner. I couldn't imagine being a
cop, working alone. When I looked back, she had a pencil behind her
ear and another in her hand, pointing at things, moving her lips
silently as if trying to tell herself the story of what had happened
that day.

"Hell of a way to celebrate freedom," Antoine said, waving his hand
to the tech as he walked back to the rig. "Hell of a Fourth of July, if you
ask me."

She smiled at us. "I'm on overtime."

By the next morning, I was exhausted. I sat at the kitchen table trying
to focus on the box scores in the paper while the oncoming shift poured
coffee and talked about the picnics and fireworks shows they'd taken
their kids to on the Fourth.

"Harpo, did you see a ghost last night? You look like shit," Jack Alvarez said as he came on duty.

"I feel like shit. Can you ride for me?"

"I've got you covered, get out of here."

I took a long shower at the firehouse, drifting asleep occasionally under the heavy jets of hot water. At home Shona was where I'd left her the day before, curled up asleep, so tiny in the big bed. She looked like she always does when she's asleep: childlike, drowsily content, and a little indistinct around the edges. I crawled in and wrapped an arm over her. She stretched without opening her eyes.

"How was your day, honey?" she said through a yawn.

"It was good. Busy. Nothing special."

"I'm glad you're home. I missed you."

"I missed you, too."

She rolled over and hugged me, burying her face in my chest. "You smell fresh."

When I woke up, it was dinnertime. There was a cool night breeze blowing through the apartment, and I could hear Shona on the balcony, tending to her plants and singing with the radio.

14

Working Fire

One day in the middle of summer—a little early for fire season but never too soon for hysteria—we got sent up to King Estates, a patch of grassland abutting a tract of expensive homes.

"I hate these damn canary suits." Jack Alvarez was my partner for the day. As usual, he was wearing his heavy cotton sweatshirt under his suit instead of a T-shirt. Even his friends, the guys who went with him on softball tournaments and duck-hunting trips, had never seen him with his sweatshirt off.

"Maybe if you'd gotten some sleep last night, you wouldn't be so grouchy this morning," I said. He'd come in looking a little green around the edges.

"I feel fine. I just haven't woken up yet," Alvarez said. His eyes were bloodshot, and he'd forgone his usual breakfast plate of reheated leftovers.

We bounced around in the back of the moving rig, struggling into our wildland gear—cheap plastic hard hats and bright yellow fire-resistant Nomex suits that are lighter and more flexible than our regular gear. Jack's jacket was filthy with black grease. He liked to wear it during his spare time at the firehouse, when he would pull out his welding stuff from under the shop bench and make gear for the weight room, or new pieces of equipment for the truck.

Cresting a hill, we could see a half acre of fire backing slowly downslope and sending up columns of smoke. We dropped a lead at the closest hydrant, and then I hopped up and gave the driver two beeps, the signal to lay line. Standing on the tailboard as we drove to the base of the fire hill, I felt like a parade of one, with all the residents of the street standing in their driveways grimly watching us pass.

Engines Eighteen and Twenty were attacking the fire from the top when we got there, so there was nothing for us to do but hike it from the bottom. I strapped on my fire shelter and a canteen, threw a two-gallon piss bag on my back, and grabbed a Macleod brush-clearing tool in my hand. I've always been skeptical of the gear that we bring with us on wildland fires. The fire shelter's name is misleading; it's really just an aluminum-foil bag that folds down to the size of a Bible and offers only slightly more protection. The theory is that when you're about to be overrun by advancing flames, you whip out the bag, hunker down in it, and count on the foil to reflect away the heat. The bags are generally referred to as "Shake 'n Bakes." Our wildland instructor in the tower had told us, "If you put yourself in a spot where you have to use it, you've already fucked up. So try not to expect too much."

California turns brown in the summertime. It is only during the long wet winter that the hills sprout bright green grass and clover. Come summer, all rain abruptly stops and the hills temper to the uniform rusty shade of parched vegetation. After the last rains in May, meteorologists don't even bother to check their rain gauges again until mid-October. When the grass dries, when the blades bend over and disintegrate in the wind, Oakland becomes a nervous city. Every thirty years or so, a major fire sweeps through the dry brush and vulnerable hillsides of Oakland's upper acreage. The most famous conflagration was in 1991, when a smoldering late-summer undergrowth fire got out of control and was whipped into a frenzy by the winds. The fire leaped from canyon to canyon, jumping an eight-lane freeway and destroying entire neighborhoods. A fire chief, a cop, and twenty-three civilians died that day, trapped in a narrow street and hemmed in on all sides by walls of flame. Three thousand of the city's priciest homes were destroyed. Since then, Oakland's rich hill dwellers have gotten to spend the summer feeling the same sort of edgy uncertainty that is a year-round condition of existence for denizens of the flatland slums.

On hot days the wealthy hang off their porches and pace their well-manicured streets, eyes scanning the hillsides and noses twitching for the hint of brush smoke. A dozen frantic phone calls will come in any-time somebody lights a barbecue in a backyard and lets a puff of white smoke drift into the air. The fire department responds aggressively to these fears. We hit grass fires with everything we've got. Tank wagons, hand crews, brush patrols, and even airdrops have become standard procedure in the summer. There has always been a lingering suspicion that with more vigilance the department could have prevented the de-struction of '91, so the current campaign-style assaults do as much for public reassurance as they do for actual extinguishment.

People hack their way into the hills to build nice houses, and the hills fight right back against them. Gutters get choked with dry leaves, and tinderlike annual grasses grow up and die in the disturbed soil that people leave behind. Years ago settlers in the Oakland area imported eucalyptus trees from Tasmania, thinking that they would be a cheap and fast-growing hardwood. The trees proved too thin and gnarly for use, though they did grow spectacularly in the cool Northern Cali-fornia environment. Unfortunately, eucalyptus has a bad habit of shed-ding branches and bark in the winter, yards-long strips of aromatic wood that in the summer lay a blanket of kindling over every unpaved inch of land. For us, the neighborhoods in the hills can no longer be thought of simply as Rockridge or Montclair. Instead we lump them all together as the Urban Wildland Interface and wait anxiously through the summer for the first relaxing rains of fall.

Since the defining feature of a wildland fire is that it's out in the boonies somewhere, it's often impossible to stretch hose lines to the fire. So we carry a few gallons of water on our backs and sparingly squirt out little drops. The hand tools are a variety of picks and rakes, all designed to clear the ground down to bare mineral soil, so there will be nothing left to burn.

I like wildland fires. The work is hard and painstaking, but it feels solid and pure. Even the smoke smells clean. It's just fire on grass, like a thousand campfires remembered from trips when I was a kid. You can almost breathe grass smoke; it burns your eyes and makes you tear, but it's subtle and forgiving, not like the panicky stab that comes with the first breath of burning plastic.

I fought a few wildland fires when I worked for the park service. When summer lightning storms streaked over the dry desert, all the rangers would get jumpy and shake out their yellows in anticipation of fat overtime checks. Fighting wildland in Oakland always makes me a little wistful for all of the time I spent living in the backcountry, sleeping on the ground. Clearing brush, I get a visceral pang and wonder whether I'll ever see clouds again, real clouds, the kind that are visible miles off on the horizon instead of the flat gray blanket that rolls into Oakland off the bay. Every grass fire on Skyline Boulevard or in the Piedmont Cemetery reminds me that I'm spending another summer in the city, sleeping in a bed and trying to pick out a star or two through the all-consuming wash of light from San Francisco.

But wildland fires have become big business in Oakland, and we aren't in a position to allow for the fact that the hills need periodic blazes to clear up the choke of downed fuel. Rich urbanites push farther and farther up into steep box canyons in search of solitude, and when the fires start in late summer, they scream for protection. The constant suppression of the little fires causes rarer but more catastrophic ones. Three hundred years ago, when the Oakland Hills were just uninhabited oak woodland, fires would sweep through periodically, clearing out dead brush and renewing the soil. But now, with homeowners leaning over their balconies and news helicopters wheeling overhead like hawks, there isn't time or inclination to consider vagaries like ecosystem health. When the chief says kill it, we kill it.

"We should just stay here by the houses and let the fire burn down to us," Alvarez muttered, shouldering his tools. He was swaying unsteadily on his feet, and I knew he wasn't looking forward to the hike. It was amusing to see Jack operating on anything less than full whirling-dervish speed, and I could tell that the shame of feeling whipped was worse than anything his sick stomach might be doing.

We started up the left flank until we came even with the fire. Another crew joined us and relayed the order to cut a line up to the dirt road at the top of the hill. Even the simplest of structure fires is exciting; small wildland fires are often notable only for their monotony. We lined up like a gaggle of convicts and started clearing brush with shovels, Pulaskis (an ax with a digging blade on the back), and Macleods (a wide hoe with a rake on the back). All the officers in their red helmets were standing around giving instructions while we whacked at the ground. "Swing loooooow, sweet chaaariooottt...," somebody started singing as we swung our tools. One step after another, a bush at a time.

"Keep one foot in the black, Unger!" a captain shouted at me. I'd bumped up from the line a few feet to root out a nice big chamise bush. Grass-fire theory says that you should never be surrounded by dry grass, should always have one foot in the relative safety of land that's already been burned over. The fire was moving slowly, the flame heights were low, and we were swarming well, like an army of termites chewing up the hillside. It was a small fire, and I'd been lax. In 1968 three Oakland firefighters had burned to death in grass no taller than their knees.

The winds were light, not the swirling, downslope "diablo" winds that can change a unit's entire approach. Grass fires naturally want to burn uphill, and we can use this predictability to our advantage, attacking from the safer downhill side. The plan was to cut a fire line along both flanks and concede the area between the top of the fire and the nonflammable road at the crest of the hill. We started our line well to the side of the fire, close enough to feel a bit of radiant heat but not so tight that the flames and smoke would impede our progress. Nobody cares how much grass we let burn, as long as the flames don't get near any structures.

At the top we met up with another crew that had been cutting line toward us. "Take a blow," said one of the captains, who was holding nothing more useful than a radio.

"Yeah, blow *this*," Alvarez said, making a halfhearted masturbatory gesture. His heart wasn't in it, but no matter how sick a fireman gets, he can't let an opportunity for a good dick joke pass him by.

"Are you going to make it?" I asked. He'd flopped himself down flat on his back in the grass, an empty water bottle in his hand. The fire skittered around harmlessly in the dry grass fifty feet away from us along the hillside. It was early enough in the season that there was still a little bit of green left in between the patches of brown.

"Wildland fires are bullshit. Gimme some water." He pointed at the regulation canteen on my belt. I turned it upside down and shook it for him. It was empty, just like his, just like everybody's. It's required gear, but they don't require us to keep them filled, and since in Oakland we're never more than a thousand yards or so from a road, nobody ever bothers to carry the extra weight.

"No luck, buddy," I said. "You look like shit. I mean, you always look like shit, but this is shitty even by your standards."

"Shut up, Harpo. I'll kick your ass in a few weeks when I feel better."

"Fair enough. Let's finish digging this line and get home."

Alvarez dragged himself to his feet and stood for a moment leaning on his knees, the fabric of his pants knotted between his fingers. Beads of sweat sprouted on the back of his neck. When he straightened himself out, we started back down the same trail we'd just cut our way up. The first pass was rough and jagged, and we finished by cutting a wide trail, spritzing water on the hot spots until we ended up back where we'd started. When you hear about western fires in the summer, the newscasters always like to talk about what percentage of the blaze is contained. All this means is how much of the line around a fire has already been dug. Once a fire is completely encircled by a fire line, it is 100 percent contained. Containment, unfortunately, is not the same thing as control. A fire that is completely contained can still be burning out of control within the fire lines. Strong winds can pick up little pieces of flaming brush and carry these burning brands over distances of up to a mile. It was the brands that turned that "nothing" fire in 1991 into the "1991 Oakland Hills firestorm." No matter how hard they fought or how many extra resources they brought in, the firefighters just couldn't move faster than the wind, and the flaming brands got ahead of them at every turn.

When Alvarez and I got down to the bottom, the fire was entirely contained. We stood for a minute, leaning on our tools, watching a few isolated bushes inside the fire line burst into flames. The engines at the top had stretched two long hose lines into the burned patch, and crews were walking bush to bush, turning over stones carefully to look for any buried hot spots. On some fiery hillsides, I've seen small logs or even cow pies roll downhill carrying the fire with them. Losing your nice little circle of containment to a rolling, flaming hunk of cow shit is about the most demoralizing thing that can happen after a long day of cutting fire line.

The wind was low, and the chiefs were satisfied, so they gave us the order to pack it in. A crew or two would mop up for an hour, and for the next twenty-four we'd make sure to have a few guys on fire watch, just sitting on the hillside to guard against a rekindle.

"Fucking wonderful," Alvarez said, looking at the thousand feet of hose we'd laid in the street. It was all heavy with water, and it went straight up and over a steep rise. He looked shaky. We trudged up the hill alongside each other, breaking the hose couplings at intervals to let the water drain out. When all the hose was dry and piled at the back of the engine, Alvarez and I climbed into the elevated hose bed and laid the line out again for the next time we'd need it.

"You all right, Jack?" I asked. I was in the cab of the engine. I'd heard him open the door for his side but hadn't seen him step in yet. I leaned over just in time to see him wiping his mouth. He was down on a knee, retching on the pavement. "How's it going, killer?" I asked, looking down on him.

"Didn't I already tell you I was going to kick your ass?"

"Easy, tiger. You're scaring me." I filed away the image of Jack reeling with nausea to be used later as ammunition for when he was feeling better. My initial encounter with him had been on one of my first fires, when he'd tricked me to get his hands on the nozzle. Somewhere over the course of my probation, I'd stopped being scared of him and started to enjoy his particular style of fire fighting. He was like a great lumbering elephant; he probably could have picked apart any problem with finesse and brains, but mostly he just enjoyed hurling himself at a fire until it was thoroughly thumped into submission. I did wish, though, that I could have had that picture of him on his knees in the gutter back when I was a jumpy new kid.

Back at the firehouse, he finally admitted defeat. It was a hot day and getting hotter; the firefight up the hill had given his flu a toehold, and there was no way he was going to be able to make it through the day. Firefighters hate to give up, because they hate to let other firefighters *see* them giving up. Leaning against the sink with his sweatshirt dripping and his chest heaving, he wasn't going to do us any good. The officer sent him home and had the battalion chief call around to see if there were any off-duty guys who wanted to come in midday for a few hours of time-and-a-half pay.

We ran with three for an hour, until the back door to the firehouse opened and Antoine Ugade showed up to finish out the shift, not as an acting lieutenant anymore, but back to being a regular firefighter like me.

"Shalom, brotherman," he said, coming through the door. He shook my hand, and we pulled together for the distinctly masculine, touch-free, backslapping kind of hug favored by firemen.

Antoine threw his gear up on the rig and double-checked his air bottle, a personal-safety ritual that none of us ever skip. I was glad to see him, since we were about to head up to the hillside for a few hours of fire watch, and the prospect of listening to the officer's aimless prattle while broiling in the sun wasn't the least bit enticing. At least Antoine would keep me laughing.

Some people come to this job because of the solid pay and steady benefits, some show up because that's what their fathers did, and some folks come to fire fighting because they've never been able to imagine themselves doing anything else. Antoine was just a natural. He'd set his sights on the fire department in high school and gotten hired by his first department when he was eighteen. "This is all I've ever done," Antoine joked. "I've never worked an honest day in my life."

Antoine's enthusiasm for fire was captivating and infectious. Sitting on the hillside, I half hoped we'd get a good fire to play with that day. It's bad form to hope for a fire, but we can't help it; fire is what we look forward to. A fire means that somebody is in danger and may lose his or her life or health. A fire means that we're putting ourselves at risk, that one of us may get dragged away to the hospital with a broken ankle or smoke inhalation or something worse. Fires are categorically a bad thing, and we wish for them every day.

Sure enough, it wasn't long before Antoine and I got our fire. And then some. The location information on the printout said Foothill and Seminary, right in the heart of the district. We were into our coats and buckled down into our air bottles before the officer even made it to the rig.

"Where the hell is he?" Antoine yelled, his legs bouncing up and down with impatience.

"It was like that this morning, too, before you came in," I said, flexing my fingers to soften up my gloves. The object of our impatience was Benedict, a captain who seemed to have no real desire to fight fire in any aggressive sort of way.

"I'll kill him if Engine Eighteen beats us into our own fire," Antoine said. Arriving second to a fire in one's own district ranks as one of the top indignities that any firefighter who cares about his image can face.

Benedict finally made it to his seat, and the engineer sent the rig flying out of the firehouse as soon as he heard the thump of the closing door. In the front seat, Benedict picked at his equipment with two fingers as if it offended him. He fumbled with the radio, trying to dial in the right frequency for the fire attack. He wasn't even close to arranged by the time we hit the intersection; his helmet was cocked off to one side of his head as if it were an ornament instead of something that might actually be useful in saving his life.

"I think we should wait here; we might want to go defensive," he said, his voice breaking as he looked up at the fire building. "Might" is a bad word for an officer to use; right or wrong, the crew is looking for some definitive orders, not a wishy-washy hemming and hawing while the whole world is going to shit.

The structure in front of us was a massive two-story apartment building on the corner of the street, and it was burning from end to end, fire blowing out every upstairs window. There was a thin alleyway separating the fire building from the as-yet-uninvolved building next to it. A doorway opened to an enclosed stairway that looked like the most obvious place to make an entrance. There was another stairway on the far end of the building, the end closest to the street corner. A high-tension power line burned up and zipped arcing through the air, just missing the roof of the engine. "Let's wait and let the chief give a good size-up when he gets here," Benedict said. Burning bits of paper were wafting down on us, and the neighbors were starting to holler. Benedict absorbed himself with getting his radio belt situated just right.

Antoine and I never heard a word he said. In Oakland it's a matter of pride not to go defensive until we've been thoroughly beaten on an interior attack. Every fire starts out small, and every small fire can be stopped, though not while standing motionless in the street. A defensive fire means setting up huge-water master streams in the street, pouring thousands of gallons into a building from the outside, drowning it from above. In a defensive fire, the building is destroyed, the floors collapse onto each other, and everything inside is reduced to rubble. The only things that are defended are the surrounding buildings.

Small suburban departments with limited experience, low manpower, and long response times tend to favor the defensive, exterior, "surround and drown" style of attack. While it's a safe approach as far as the firefighter is concerned, every fire that is fought defensively means that a building has been reduced to a parking lot. Big-city firemen want to go inside, to feel the heat curling the backs of our ears and stay inside fighting fire until the last possible moment, when it's clear that a save is impossible. And then just one minute beyond that; we hate to build parking lots. Big-city firefighters hate defensive fires because there's nothing to do but stand around and watch as the building goes through its death throes. Going defensive is giving up.

Tom once told me that when he wasn't working with Evan Powell, he "reserved the right to ditch any officer at any time." He told me not to trust someone just because he's wearing a red helmet. "There are a lot of officers who'll get you killed if you're not careful," he'd said. Benedict wasn't going to get us killed, but it didn't look as if he would find the ability to put himself into motion either.

Leaving Benedict and heading inside was on the thin edge of insubordination, but neither Antoine nor I considered staying with him. As long as we didn't get hurt, there wouldn't be any problem, and neither of us was planning on getting hurt. Besides, Benedict had a reputation, and whatever battalion chief showed up would understand why we didn't stick around with a bum captain.

Antoine pulled a preconnect attack line off the rig and headed for the fire. I grabbed another line and followed him. It was looking like one of those nights where there wasn't any need to fight over the nozzle, because there would be plenty of fire for everyone.

I opened the nozzle on the alleyway between the building on the corner and the next apartment complex down the block. Flames were breaking through the window of the first building and leaping through the air into the second. I knocked back the flames, trying to keep the fire from spreading to the second building, but when I momentarily shut down the water, it was clear that we were far too late for that. Flames shot out of the window of the second building as soon as the water was turned away; the fire was spreading, and there was no use in trying to protect one disaster from the other.

Another engine crew arrived, and people spilled out, eager to help. Nobody was smiling exactly, but everybody was happy to be there.

"I'll go down the block!" Antoine shouted at me, pointing to the second stairway down at the far end. He grabbed someone, and I saw them disappear through the doorway. Another firefighter peeled off and picked up a spot behind me on my hose line so we could start up the stairs of the second building. With masks and coats and helmets, everybody looks pretty much the same, but even if you don't know who's working with you, you can tell if they're going to be good teammates by the way they move, by how eager and aggressive they are in pulling line off the rig. In general, I don't like splitting up the crew, but since Benedict still wasn't doing a damn thing and there was plenty of help around, it felt like our best choice. Antoine led a makeshift crew up one hot building, and I started into the other. It was just confusion, but everyone knew what to do and silently went to work. It was the kind of confusion we could handle.

We—myself and whoever it was that had fallen in behind—were in some sort of enclosed exterior stairway, like the common entrance to an apartment. The people running out had left the door open, and we had a clear shot into the fire room. I lay on the ground on the doorstep, firing water into the room. It was smoky, but the fire lit the place up a little, and I could see lots of chairs and a table, a bookshelf that had fallen over and cantilevered against an opposite wall.

"Heads up," the guy behind me said, or something like that. It was hard to hear over the water and the fire and the sirens and the chain saws working away on the roof. Everyone's voice sounds the same through an air mask.

I rolled over onto my back and looked up to where he was pointing, to the window on the other side of the landing.

"Hit *that!*" he yelled, and that time I could make out the words perfectly even through the mask, because his voice was loud and urgent and I was thinking exactly the same thing. Flames were coming through the window opposite the doorway where I'd had all my attention focused. It's a textbook bad situation—fire on both sides—and the flames were trying to meet in the middle, which happened to be where we were kneeling. I was worried that through the window was another apartment, maybe even another entire building, that we were in the process of losing. I knew that Antoine and whomever he'd picked up were fighting fire in the building on the far side of my big fire room, and I could only hope that the battalion chief in the street had called a second alarm, a third, a fourth, so that there'd be someone fighting fire in this new building as well. I don't love having two fires to fight at once—it sets off those little alarm bells in my head—but we were right up a straight stairway and could always bail if things went to shit, just roll down the twelve or twenty steps and be on the curb in a matter of seconds. There wasn't any reason to leave yet, though there's always plenty of reason to start getting nervous when there's fire on two sides and only water enough for one.

For a few minutes, I alternated spraying water into the room with knocking back the stuff coming in through the window on the other side. Each fire got smaller for a second and then flared up again when I switched sides. It didn't feel as if we were in any imminent danger, but we weren't doing much good either, just treading hot water.

"Let me have it. I've got an angle." The firefighter I was with came up to a half hunker, and I handed him the nozzle when he reached out. Then, like a gunslinger from an old western who was hopelessly surrounded, he stood up quickly in the window, fired out through it for a few seconds, and dove back down into a crouch. When he went up again and punched out the window frame, I could see the name on the coat: D. KINNEAR, which meant the mystery man wasn't a mystery—or even a man—at all but Deb, one of my classmates from the tower. If Deb were a man, everyone would say that she's "a helluva firefighter," but since she's not, the old guys aren't able to bring themselves to give her that courtesy. But she's got a nose for fire, good instincts, strong moves. At first I was a little worried when I saw her, because we were without an officer and she didn't have any more experience than I did, except for a few years at some hick cow-town fire department in the foothills. But what the hell? We'd both seen plenty of fire by that point, and nobody was going to put the fire out if we didn't. All of a sudden it started feeling good that it was us, two new kids together, taking a beating and doing our best. I was glad it was Deb with me on that landing.

To buy herself a second, Deb knocked back the flames licking out from the main room, then stood up and leaned out the window where that other fire had been coming from. Down on the floor, I fed her line and could hear her spraying.

"It's just the siding," Deb said, slumping back down onto the floor next to me. She lifted her face mask a little to let me hear her better.

"It's our building, just the outside wall. No problem." She stood again and jumped up so she was leaning out of the window from the waist. I took the weight off the line for her while she sprayed down and up and along the outside walls, allowing us to concentrate on the bigger battle waiting inside.

Meanwhile, without getting the proper attention, the flames in the main apartment were becoming greedy, moving right back where they'd started, filling the room, punching out through the doorway and over our heads. For the first time, the word "defensive" was starting to enter my head. But the heat wasn't *too* bad, the stairway still felt reasonably solid under my feet, and nobody's bell was ringing on the air bottle yet. I lay to the side of the doorway, leaning around to get the nozzle in place. It was kind of a lazy feeling, like being in a trench waiting for the attack. We weren't working, exactly; spraying water doesn't take much physical effort. Every now and then, I pulled up to my knees and tried to move deeper into the room, but there was too much heat and flame. It was starting to look as if the fire was burning from below as well, flames coming up through the floor, and I didn't want to go in there just so I could fall through into the burning street-level unit. So instead we held the line; we weren't making any headway, but the fire wasn't either. I whipped the nozzle around and around in swooping circles, letting the water hit the ceiling and shatter into millions of tiny droplets. Water doesn't drown a fire so much as it distracts it. The fire has to expend energy to turn all that water into steam, and our whole game is to get the fire so busy making steam that it doesn't have any strength left to do anything else, like burn through the floor.

We continued to hold the line, lying on the floor and shooting water. It can get a little boring, actually; it's easy to get distracted, and I was letting my mind wander. I tried to make out the titles of the books on the floor across the room; I stretched my legs and rolled around to get in a more comfortable position; I watched the smoke rolling lower and lower down the walls, banking like bay fog. The diesel engines in the street and the saws on the roof were blending their sounds together into a distant hum, like white noise—grating if you concentrate on it, but somehow soothing when you let it just wash over you. When you're inside a house and it's just you and the fire, crouching together in a sweaty détente, the building crackles like a campfire, and flaming ceiling tiles and doorframes are almost beautiful as they fall from above. It's safe and clean inside the mask; your own breathing is loud and raspy, reassuring because it's so obviously *there*, uninterruptible. You're kneeling on the floor, gathering up before it's safe to go in; it's hot but not unbearable, dangerous but not scary, work but not labor. At times like this, the rough metaphors about "battling the beast" and "slaying the red dragon" don't fit; fire can be quiet, contemplative, a melancholy blue.

I felt Deb's hand snaking over my shoulder and reaching for the nozzle. She put her hand over mine and whipped the nozzle around violently, spreading the water throughout the room. And suddenly I was embarrassed; I'd been admiring the fire, enjoying the moment, losing myself in a reverie. Deb wasn't trying to steal the nozzle from me—she had too much poise and self-confidence for that sort of game—but she was trying to snap me out of it, bring me back down to the fact that the *shit is still burning,* and we're *firefighters,* and if I can't move inside yet because it's still too hot and I don't trust the floor, then the very least I can do is pay attention and whip the nozzle around and make my best effort.

The muffled chorus of mechanical grumbling that had been lulling me finally built to a crash, and a big patch of ceiling fell down, hinging off one side like a massive trapdoor. Up on the roof, the truck crew must have been backing away from their good cuts. Once the roof was open, the hole would be spewing flames and heat and all the gas and smoke trapped inside that had kept us from advancing. The smoke lifted somewhat, and I could see the floor, a bit of wall as high as the electrical outlets, then the bottoms of the windows. The heat lifted some more, and for the first time I made the corner, entered the room, stamping my feet in front of me with every step to make sure the floor-boards hadn't gone spongy with fire from below.

But it all felt solid, and unconsciously I stood up straighter and straighter, because the heat was leaving and we were winning. The fire was going out. I knocked all the flames in the main room; the blaze was defanged and almost vulnerable now that the roof was opened and all of the fire's strength had drifted through the hole and up into the night sky. We moved into the other rooms, still smoky but cooler now, the small lamps on our coats making little cones of shaky, dusty light in front of us. Deb was off in front of me on all fours, doing a search. I couldn't see her, but I knew she'd be down low, keeping her right hand on the wall to stay oriented, making left turns at every corner, check-ing under chairs and swinging her legs out into the middle of the room searching for bodies.

The back room was unventilated, still smoky and wet-hot. In the dark I flipped over a mattress, pulled a dresser away from a wall, rum-maged through the debris in the bottom of a closet. When the smoke cleared, I'd do a more thorough secondary search, but for the moment I just wanted to get a good rough idea that there was no one in there but me. The back window had somehow managed to stay unbroken, but the plastic latch was melted. I picked up a clock radio from the floor and used it to smash the window glass out of the frame, then opened my nozzle and aimed a stream out the window. As the water flowed, it made a whirlwind of air that cleared the room of the smoke and the soot and the darkness.

One of the guilty pleasures of fire fighting is the thrill of destruction. It's every boy's dream: I destroy things for a living. Need to clear the smoke? Smash a couple of skylights with an ax. Table in your way? Flip it over and throw it out the window. Every potentially affected space has to be opened and overhauled; walls and ceilings are methodically chopped apart and dumped onto the sidewalk. Often the entire roof has to be stripped, completely dismantled by a firefighter with a chain saw crawling along the steep pitch of a slick roof that is loose and thin from the flames that burned beneath it. The perverse thing is that fighting fire is fun. Residents are standing in the street watching everything they own be destroyed, and we're having the time of our lives. Fire is inevitable—somebody might as well enjoy it.

"All clear?" Deb poked her head into my room, and I nodded. She leaned out the window and shouted, "Secondary search clear!" down to a red hat standing in the street. I heard "All clear" repeated in stereo on all the radios on belts and in rigs up and down the street. It meant that there was nobody but us in there, that whether the building was going to end up as a total loss or not, Deb and I had fought the fire until we won and we didn't let anybody die while we were doing it.

Deb had her face piece hanging around her neck; it was still smoky, but it felt nice to get out of the mask, so I pulled mine off, too, and we stood in the middle of the wreckage together for a minute. The hard work was just about to begin, the overhaul, the backbreaking search for hidden fire, but there was time to stop to talk.

Deb doesn't have it easy. It doesn't seem like fun to be a woman in the fire department. I can't even count the old guys with giant bellies and busted knees who sit around all day grousing about how no woman could possibly do the job. Deb doesn't let it bother her, though; she just goes out and works, fights for the nozzle, pulls ceilings tirelessly. She's a natural, and she doesn't care if anybody else admits it or not.

"I haven't seen you in months. How you been, Unger?" Deb was the only person who actually knew me and still called me that, just "Unger." It always struck me as a little funny and formal coming from somebody whom I think of as my friend, even if we never see each other.

"Great, great. Hanging in there. How's Jim?" I asked.

"He's good. How's Shona?"

"She's good, too." I've never met Jim, and she's never met Shona, but it's the thing to say. And then we asked, "Kids yet?" and "Buy a house yet?" and it's no all around.

"Are you still at Station Eighteen?" I asked, kicking at my boots as if the ash were clumps of dirt I could just shake off.

"Yeah. You still downtown?"

"Yeah. That's my spot." And that was about all there was time for before it felt as if we'd been talking too long and needed to get back to work. I enjoy all the bullshitting that happens when firefighters from far off get together at a big fire, but I always feel guilty, like I'm stealing time from the city and we should hurry up and get back in service.

From the street a heavyset black man threaded his way past the engines and approached me. He was wearing shorts and flip-flops, as if he'd just been awakened from sleep.

"Hey, man, is there anything left up there?" he asked.

"Not much."

"I got some tools in the back closet." He pointed to the blackened window frame of one of the first rooms Deb and I had made it into. "I have to get those tools. Could you look for me?"

I went back upstairs and rooted around in the closet underneath a soggy pile of clothing. The heavy metal box was wedged under something, so I kicked it hard and then set it on the ruined bed. A cordless-drill set, a Skill saw, and a nail gun were all melted together in a solid mass of plastic-encased metal. I opened the latch on the toolbox and poured off all the water that had leaked in through the edges.

"Here you go," I said when I got back downstairs. "The rest wasn't worth saving."

"That's all right, man," he answered. "I got the basics right here. As long as I got these, I still got a job." The windows of his apartment were puffing smoke, and firefighters were throwing everything he owned out the window. He opened the toolbox and cradled a pipe wrench as though it were an infant. "I don't know how to thank you."

"I just wish you didn't have to," I said, wondering why I'm always re-duced to platitudes whenever I talk to a citizen. "We did everything we could, but the fire was already in the walls before we got here. I'm sorry we couldn't save the rest of your stuff." It felt like a false thing to say, but it's really what people want to hear.

"You guys are my heroes, man. No bullshit. You guys are the best." In the real world, people never give blanket compliments like that. Firefighters hear them all the time. The praise made me uncomfortable. He wrapped me up in a giant bear hug and kicked his toolbox for em-phasis. "Thank you, brother. God bless."

When I stepped back from him and turned around, I saw Antoine, who was standing at the curb watching the whole scene. I hadn't seen him since we'd both jumped off the rig and abandoned Benedict.

"So you're a brother now?" He grinned. "I didn't know a fire could do that."

"How'd you do, man? Get some good fire?"

"The best. That was fucking great. I think *he* wants to talk to us, though." He looked over toward the rig. Benedict was standing beside it, clean and dry as a bone. It didn't appear he'd moved since the be-ginning of the fire.

"Wonderful. Let's get this over with."

Aggressive attack or not, the building looked like a complete loss. The exterior walls would make it, and maybe a bit of the floor, but be-yond that, everything was garbage. Standing in front of Benedict, I crossed my legs to try to hide the giant tear in the crotch of my slacks. One of the disadvantages of being a fireman with no ass whatsoever is that the combined weight of my ax belt and a few gallons of water al-ways pulls my pants down around my hips so that they invariably tear when I try to go up stairs. I pretty much destroy a pair at every fire.

I looked up at the building while Benedict talked, giving his oblig-atory officer postmortem. He was making some scolding noises about sticking together, but it was clear that he was mostly just relieved because we'd made a good stop, nobody got hurt, and he didn't have to go inside. We made him look good for the battalion chiefs. Antoine was nodding, looking attentive, but when Benedict bent over to get something out of his shoe, Antoine kicked me in the foot and rolled his eyes. We hung our heads in apology and muttered a few reassurances that nothing like this would ever happen again. I could tell that Benedict didn't care, but he felt as though he had to make a statement to cover his own ass in case somebody asked why he didn't have a dirty spot on him. I missed Captain Powell's calm assurance on the fireground, the feeling I always get with him that the situation is under control and I will be looked after. But it was exciting, too, to be on my own, to have had enough confidence to know that my officer of the day was lost and that my skills had grown to a place where I could begin to rely on myself.

I went into Antoine's side of the building, and he told me about his "cocksucker of a firefight" and I told him about mine. We both had pike poles in our hands and were punching holes in the ceiling, working the hooks between the slats of lath and pulling down great hunks of plaster and insulation. I poked at the walls with an ax, stripping everything down to the rafters until the whole room was just naked wood. It was like standing inside the rib cage of a decaying beached whale.

"This room's not going anywhere," Antoine said, and he moved quickly to the next, peeling away the edge of a doorframe, a favorite place for fire to hide. Every good fire involves a few minutes of fun—kicking down doors, crawling under heat and smoke, the chance to open the nozzle—and three hours of exhausting, boring, carcinogenic toil. They didn't say a thing about that in the tower; it wasn't on the bus-bench billboard either. But it's what we do, and nobody complains much, not seriously anyhow. Because it's quicker just to work fast and quietly and try to get back to the firehouse as soon as possible for a shower, some ice cream, and, if we're lucky, the last few innings of the A's game on television.

After a good fire, there's a feeling of satiation, and the firehouse looks a little different than when you left it. If you were in the middle of cleaning behind the fridge when the fire hit off, you come back afterward, look at the half-clean spot, and leave it for the next day. After a few weeks with no fires, it gets easy to be lulled into thinking that we're janitors, that keeping the firehouse clean and grousing about the administration are the reasons we're there in the first place. After a fire everything resumes its proper significance. We've been doing what we're paid to do, the thing we love. Suddenly a little dirt behind the stove or an inane new policy from upstairs doesn't seem like something to waste good energy worrying about.

Five hours after the bells went off, after our legs had cramped, shoulders stiffened, and eyes gone red and watery with fiberglass dust, we dragged ourselves back into the firehouse, freezing and triumphant.

Antoine pulled the dirty tools off of the rig, the axes and Halligan bars and pike poles that we'd used for overhaul. He wiped them clean of soot and plaster dust, using a rag to work clean oil onto the metal so it wouldn't rust. I took the face pieces off the masks (except for Benedict's, which was still clean), washed them so we could see, and shook out the dust so we wouldn't have to breathe it next time. The rig was filthy with ashfall; we'd parked right in front of the fire building, and every bit of dirt and debris and floating wreckage had ended up on the roof, the windows, the back step. I wiped the mirrors and the windows and made a brief attempt at the smudgy fingerprints by the door, but it was late, and we were too tired and hungry and excited to do an honest job of washing the whole rig down. I was shivering in my soaked T-shirt, and my torn slacks were hanging down like a loincloth.

I was ready for the night to be over, but Antoine was just cruising, working on the tools, whistling—actually *whistling*—while he worked. I wondered where he got his energy from when I was so close to falling over with exhaustion.

"I'll be in the shower," I called to Antoine, who was giving the rig a final once-over to make sure that everything was in place.

"What? Do you want me to soap your back?"

"No, just give me a holler if we get something."

"I got you, buddy. Hurry up so I can get one in, too."

The shower felt hot and warm and strong. Somehow the firehouse avoided the California mandate for anemic low-flow showerheads, and I have always enjoyed the wasteful gallons of water without guilt. In general, I try to keep my showers short, because the bells are hard to hear under the running water, and I hate the scramble of diving into my clothes soaking wet when an alarm comes in.

Antoine came running into the bathroom. "What's up?" I asked, toweling off.

"They've got something going on West Grand," he said, trying to get in one last trip to the urinal before we got called out. Going on a run with a full bladder is even worse than being called away from the shower.

I dressed in a hurry, pulling on a fresh T-shirt and giving up on my ruined slacks in favor of my heavy, fire-resistant bunker pants. Everyone was gathered in the watch room, listening to the radio, looking at the dispatch computer, and watching CNN, which for some reason had better footage of the fire than any of the local stations did.

As we drove down the freeway, the glow of an entire neighborhood burning was visible from miles away. I've had a few guys tell me that the most fun they've ever had in the fire department was the night of the Rodney King verdict, when everything was burning and all they could hope to do was darken down the flames in one house before flying off to an even bigger fire down the street. I was wondering if this was going to be my Rodney King night, or whether it would be like the '91 firestorm—scary as hell and nowhere close to fun, because the whole world was collapsing and the fire wasn't obeying any of the rules about when a fire is supposed to go out.

We were the last engine to arrive on the sixth alarm, and we were coming from far away, which meant that we had a long time to get ready, to adjust and readjust the straps on the air bottle, to think about what we were in for. The ride seemed endless. But the waiting wasn't the old fear anymore so much as it was impatience, excitement, the thrill of a kid's getting to tear into all of his birthday presents at once. It's a tremendous feeling, like *I'm good and strong and competent*, like *I'm skilled and worthy and brave*. It's knowing that I'd never fought fire with Antoine before but I'd be right there with him, I'd stand in the exact damn center of the worst inferno either of us has ever seen, and I'd watch his back and trust in him unflinchingly to watch mine.

I love these relentless nights, when good is unequivocally being done, when I feel strong and sparkling like my Grandpa Charlie and wise and kind like my Grandma Carol. And I love the mornings after a long night fighting fire, the fresh perspective on the dawn that can be earned only by not having slept during the long night leading up to it. I love pulling up at home on our safe, dead-end street and feeling the tiredness in my limbs as I unfold myself from my truck.

After a few years in the department, Shona and I have both been lulled by the routine of my disappearance every third day and by my safe return twenty-four hours later. For the most part, she doesn't ask about the scary stuff, and I don't tell her. "We had a decent fire last night," I'll say, but leave off how it feels to have flames curling over my head. She knows how dangerous my work is, but we're both happier when we don't focus on it. If I'm lucky and my relief man shows up a little early, I can get home and catch her in the moments before she leaves for work. I love to give her a strong, enveloping hug as she's walking out and I'm walking in, and then to be aware of nothing else until she kisses me awake again in the evening after her own long day of work is through.

Antoine was sitting in the seat across from me nodding his head and tapping his feet as if he were listening to music, only there wasn't anything but the sound of the siren. He was feeling the excitement, too. "Look at that," he said, smiling, pointing out his window to where the glow was differentiating into individual flames leaping off half a dozen buildings.

We screamed down the exit ramp and sped through empty streets, the smell of fire in the air, or maybe just in my soaking-wet dirty coat, my poorly washed hair, and the streaks of black soot still filling the creases of my face and neck from the last fire despite my halfhearted attempt at a shower. The otherworldliness of it was magnificent, the way the city was deserted and in flames and radios were crackling, people calling on us to help because we are the last and only option before chaos takes over. It's the closest I'll ever come to my childhood dreams of polar rescue or summiting Everest or making landfall on a life raft. As it turns out, I didn't get very far; we were rounding a corner just a mile and a half from where my parents were asleep in the house I grew up in.

A fire this massive has discrete fronts, like battle lines. No single person can control the entire zone of engagement, so the chiefs organize their troops into a hierarchical command structure, assigning strike teams, task forces, geographic and functional responsibilities. Coming in, we heard that we'd be assigned to "Division Twenty-one," or Twenty-first Street, blocks away from the origin of the fire but at this point one of several epicenters of destruction.

"Holy shit," Antoine breathed out, slapping his hands together. The collar of his coat was down and lay smooth over his shoulders. I wondered if I looked that natural in my costume. I turned my head toward his window just in time to look down West Grand as several stories of a flaming wooden skeleton collapsed with a suck of air I could feel even over the wind speed of driving. It wasn't our part of the fire. We flew past without a thought of stopping.

Every hydrant in the neighborhood had already been hooked and was flowing water. They didn't call us for our rig; they called us for our manpower. We parked a block away from our own small battle, and Antoine and I both shouldered our loads—hose-extension packs, pike poles, axes, anything the crews at the front might need.

The whole burning neighborhood called for a delicate admixture of aggressive and defensive tactics. The wind was still whipping violently, carrying burning shingles from one house to the next like a fast-running brushfire. We didn't have the resources to fight fire in every house the way we would if there was only a lone structure burning. The command staff wanted us to establish a firm perimeter against further spread while at the same time saving all the houses within the ring. It's a mistake for a firefighter to think about strategy on that large a scale, though; we had a single three-story Victorian in front of us, it was going to the moon, and we'd just have to trust in the white-helmeted chiefs to look out for the large problems.

"Pull me up a big line. Pull me up a big line, goddamn it." An officer I'd never seen before was standing next to an engine and screaming out to nobody in particular. He was probably one of those guys who'd transferred up to a quiet spot in the hills ten years ago and hadn't fought a fire since. "Who are you two with?" he said, wheeling around, catching sight of Antoine and me.

"We're with you, sir," Antoine said, calm, even amused at the eye-popping spectacle in front of us. From the moment we hit the street for this latest fire, there was no question of staying with Benedict. With the first fire of the evening, the apartment on Foothill, he'd ceded his authority to us, wordlessly given us our head to do whatever we thought safe and necessary.

"All right, then," the officer said, seeming surprised that somebody was apparently going to obey an order of his. "Get me a big line to the back window." The "big line" is just that—a giant, heavy hose. Big fire, big water. Unlike our regular attack lines, the big line is generally too heavy to be used on an interior operation; it's for defensive fires, when the better part of valor is to destroy a building, kill one to save a thousand.

"I'll flake it. Go on," Antoine said to me. I put a few feet of hose across my chest like a bandolier, grabbed the massive nozzle in my hand, and set off down the alleyway at a trot. The hose grew heavier behind me as more of it fell from the engine and dragged on the ground. By the time I reached the back of the house, it was like running through molasses, and I was pumping my legs at a furious run and gaining only inches at a time. When somebody lightened the line behind me to help, I fell flat on my face, but in the general air of confusion, no one noticed or cared.

I pulled the hose into place, then laid a wide loop into the end of it so that it crossed back over itself like a cursive e. The big line packs a punch, and I knew we'd be there for hours. When the water shoots out, the force it exerts backward is tremendous. I set it so that instead of pushing back on me at the nozzle, the pressure would be dissipated into the stabilizing loop of hose on the ground.

"Water!" I called, swooping my arm over my head. I couldn't see the engineer far back in the darkness somewhere, but Antoine was there, my cutoff man, and he relayed the signal. I watched the hose plump and slowly fill, like a snake swallowing a mouse, until the first pressure hit my end. I checked for the tenth time to make sure the nozzle was closed and waited for all the water to arrive. When it was time, both Antoine and I took long pieces of webbing and built little straps to help us handle the weight and force of the nozzle reaction. Antoine stood behind me, his foot against mine, his shoulder against my back. We were twenty feet away from the fire, set up in a side yard, and the radiant heat was blistering. Fire sprouted from all the upstairs windows, licking into the eaves, washing over the roof when the wind shifted.

"Ready for water?" I asked Antoine.

"Damn straight." It was like Hollywood dialogue, but then again, everything around us was like a massive special effect. I'd never talk that way at home, but on the fireground it just feels natural to be John Wayne.

I opened the nozzle, and we both leaned into it, the heat of the fire momentarily cooled by the waves of mist the wind blew back onto us from our own hose stream. The water from the big line did nothing; fire kept roaring out from the windows. I spent a few minutes on each window, trying to arch the hose line just so in order to plop the water perfectly through the hole. It's like a carnival game, only we were wetter, colder, more miserable. Occasionally we'd get a giant blast of water in the face from whoever was working his own big line on the opposite side of the house. We were probably giving them the same.

"We're done, then, huh," Antoine said, hitching up his hose strap as fire roared into the black sky overhead.

"Yup," I answered. "We're planted." That was it for us, our spot for the rest of the night. Baby-sitting a big line. We were just tools now, our bodies important because they're the best thing that anybody's come up with to hold down a big line. There's nothing about this that's smart or brave or fun. It's just work, but that's okay, too. I was exhausted from the day, but it was my mind, not my body, that needed relief. I was happy to be nothing more than a pair of arms, happy to take a pounding from the constant mule kick of the water. It seemed like years ago, or at least the shift before, that I was at a grass fire, watching Alvarez go belly up. Antoine and I would call him in the morning to heckle him, and he would be livid about all the action he'd missed.

When the dawn finally broke, the entire neighborhood was steaming like a bomb crater. Twelve houses had been completely destroyed, and many more were severely damaged. In all, 250 firefighters came to the scene, and we called in help from every surrounding jurisdiction to cover the normal stuff, the heart attacks and the trash-can fires, while we were all down on West Grand. The oncoming shift would be there all day, sifting through the rubble in search of hot spots and walls that were threatening to collapse.

I was shivering and exhausted, soaked through to my skin from the sweat and the water and the exertion. Antoine sat on the tailboard wringing out his socks, pouring dirty water from his boots. "Great shift," he said, clenching back against chattering teeth. "Great fucking shift."

In the rig Antoine sat opposite me and leaned his head back against the seat, dozing as we rolled along the freeway. It was the first time I'd ever seen him look less than fully alert. He was just then letting himself slow down, letting his muscles relax. I'd been feeling dead tired since sometime in the middle of that apartment fire so many hours ago; since then I'd just been feeding off the fumes of Antoine's energy to keep me going.

Antoine opened his eyes, rolled his head toward me. "We ever find out how any of those fires started?"

"I didn't," I said.

Antoine shrugged, satisfied with my answer. Then he turned back toward the window and closed his eyes again. The handle for my window was broken, and I couldn't shut it. I put my head down in a weak attempt to block the wind from whipping through my clothes. My fingers were cramped from holding the big line, my back hurt, my lungs ached from all the smoke, and my eyes were red and raw. It felt great.

15

Three Bells

On a day off for me, a brilliant, cold January morning, I went for a run in the hills that overlook Oakland. On that run I noticed for the first time that Oakland felt like *my* city. Not just a city I'd grown up in but a city I belonged to and, moreover, a city that counted on me to look after it. Running through the neighborhoods that had been burned over in the firestorm a decade earlier, I noticed that for the first time I didn't feel like a foreigner in my own home.

From my vantage along the ridgeline, I could see that, for all its problems, Oakland is a beautiful city. Standing at the pinnacle of the hill, I looked down at wide swaths of green space as Oakland tumbled, sparkling, toward the bay. The fog was holding steady in the distance, blanketing San Francisco with dark and cold, but Oakland was radiant with sunlight. Lake Merritt, with its long arms and wide inlets, sprawled out like a cat catching a nap in the sun. Growing up at the very tip of North Oakland, I had never felt any particular connection to the city beyond a zip code, but now, every bit of Oakland held meaning for me. I'd restarted a woman's heart on the floor of the Federal

Building; I'd burned my ears at a fire on the top floor of the Kaiser Building; I'd pulled a broken body from under the rails of a subway car just past Fruitvale Station.

I ran home feeling light and strong. Stretching after the run, I leaned against my truck and noticed a piece of notepaper tucked under the windshield wiper. The handwriting was instantly familiar, belonging to an old girlfriend whom I hadn't seen or spoken to in years. It said, "They didn't say a name on the news and I was just so worried about you I had to come by and make sure your car was here. I'm glad you're OK and so sorry for your loss." I didn't know what she meant, didn't want to let myself guess. My answering machine was alive with blinking lights. Some network had carried the story nationwide: OAK-LAND FIREMAN DIES IN BLAZE, TWO OTHERS CRITICALLY WOUNDED. Friends, relatives, people from my distant past, everyone who had been so excited for me when I got the job, had called and left hesitant messages. "They said he was a veteran, so I know it's not you," and "I heard it was near Auto Row, and I know you work downtown, don't you?"

I felt sick to my stomach and light-headed, but also strangely detached. I put off calling the firehouse. If I just didn't call, everyone would be all right. I took a long, slow shower, dressed, and fixed lunch. I sat on my balcony and ate, looking at the hills I had just come out of and thinking of all the people I'd met since my first day in the drill tower. My classmates, my friends, people I'd already become closer to than I'd realized up till that moment.

Finally I called. Antoine answered. His voice was empty of his usual humor.

"Antoine . . . who was it?"

"Tracy, Zac. Tracy Toomey. He's an old guy, I've never met him. There's two others down, and we don't know anything about them yet."

"Be careful, Antoine," I said, almost choking. "I'll see you in the morning."

The man who took me to my first fire was dead. He was crushed under a heavy wooden beam in a nondescript old building sandwiched between auto repair shops on Broadway. He'd volunteered for a shift of overtime, and, with the same enthusiasm he had when he was my age, he answered the bell summoning him to his last working fire. I hadn't known Tracy well at all, spent only six hours of his fifty-two years alongside him, but I was deeply shaken. Until then I had given lip service to the dangers of fire fighting, brushed aside the worries of my mother and my wife.

I didn't learn all the details until later. The crew had done a heroic job trying to save Tracy and the two others trapped beside him under the rubble. With fire still rolling overhead and water from the ladder pipes rapidly rising and threatening to drown victims and rescuers both, the crew of Station One—my station—had hurled themselves at the rescue. The rig was stripped bare of tools; every lifting bag, hydraulic cutter, pry bar, and gloved hand was strained to its limit. At great risk to themselves, as the fire stubbornly refused to die, the crew had extricated the two wounded. And Tracy's body.

I was on duty the day of Tracy's funeral. I did my housework listlessly, wishing I could join the caravan of fire engines that was scheduled to wind through the streets of Oakland. Just minutes before the procession was to start, there was a ring on the firehouse doorbell, and I looked out to see a fire engine with four firefighters standing at attention in front of it. They were off-duty firemen from a distant suburban department who had borrowed a spare rig and, on their own time, come to Oakland to relieve us so we could pay our last respects.

At the starting point, I saw dozens of fire engines, dispatched from all over the state, there to send Tracy off with honor. Two ladder trucks were parked nose to nose, their aerials fully extended and a massive American flag hanging gracefully between them. A police helicopter circled overhead. All around me were firefighters, most of whom had never met Tracy yet felt the loss as viscerally as if he had been one of their own.

Oakland is a long, slender city. Fourteenth Street, Oakland's main artery, runs the entire length of town. It was a natural choice for the procession, and the line of fire engines made its way down the street slowly, silently. It was surreal to see lights flashing all around me yet hear no sirens. We began in West Oakland, once the province of the city's elite, now home to Victorian gingerbread houses slowly sinking into disrepair. Kids on their bikes stopped and stared as we went by. An entire junior-high-school class was lined up on the sidewalk. Some of the kids were crying. Every street corner we passed was blocked by a fire engine, lights flashing silently, the crew standing alongside. Every overpass we crossed under was topped by a fire truck, its aerial extended straight up to the gray sky.

We passed through Oakland's bustling high-rise district at midday. Busy executives set down their briefcases, business meetings forgotten. Shopkeepers emerged from behind lunch counters, wiping their hands on their aprons, and stood shoulder to shoulder with bums and businessmen, all transfixed by our passage. I saw a homeless alcoholic, a regular patient of ours, standing motionless and proud, snapping off a crisp salute, dredging up through a cloudy mind a picture of himself when he was still young and proud, when he, too, had served his country.

For once the streets of Chinatown and Little Vietnam were empty. No double-parked delivery vans blocked our way, and we drove straddling the center yellow line. Children pressed up against the chain-link fences of their schoolyards to watch us pass. In the Fruitvale district, the center of Oakland's vibrant Hispanic community, the scene was the same. The roving tamale salesmen brought their carts to a halt, women kissed the babies in their arms, and the men made eye contact with us, nodding quietly. The procession finished in East Oakland, a predominantly African American community characterized by tiny houses with clean, perfectly manicured, postage-stamp lawns. Old men waved flags from their porches. The day existed out of time; the normal bustle of Oakland had drifted to a halt.

Oakland likes to claim that it is the most diverse city in the country. But that day I saw no diversity whatsoever. Only people, all coming together to acknowledge our sacrifice, each of them aware that, if called upon, any one of us might fall in an attempt to save any one of them.

As a kid I had always dreamed of being on the floor of the Oakland Coliseum, but this was far from what I had imagined. In my mind I had seen myself stealing the ball from Magic Johnson, tossing an outlet pass to Sleepy Floyd or World B. Free, being fed back the ball, and dunking to the screams of the capacity Warriors crowd. Now, for the first time, I found myself on the floor of the arena, the baskets retracted far into a darkened dome. I was surrounded by firefighters. Everyone I had ever worked with crammed the seats around me. I saw Captain Powell across the room but couldn't make my way through the crowd to talk to him. He looked good in his dress uniform, uncomfortably good for something that he shouldn't ever have had to wear in grief. His face was sad but oddly placid, as if he'd been to too many of these firefighter funerals in the past and knew that he'd go to too many more in the future.

The stadium seats were filled with firefighters from distant jurisdictions. Firefighter unions from as far away as New York had sent delegations to honor a fallen brother. Wives and children of firefighters sat in the stands, watching these strong men struggle to hold themselves together. The mayor and city-council members walked in, looking haggard and drawn, performing the worst function of public life. It had been a difficult week all around—an Oakland cop was killed the same day as Tracy, felled by a bullet from a sniper who had lain in wait on a freeway overpass.

When the lieutenant who had been trapped inside with Tracy came down the aisle, an already quiet house became absolutely silent. His eyes were red and swollen, and his best uniform pants were split clear up to his hip to accommodate his cast and the dressings on his badly burned legs. He was being supported—almost carried, really—by two men from his crew, and when they deposited him in the front row, he looked exhausted and small. I looked across the aisle at Tom McFarlane. He was crying, and making no effort to hide it. I don't remember much of the ceremony. Lots of speeches and tears, bagpipes. A bell that was rung three times in the traditional send-off for a fallen fireman. A military honor guard cutting crisp corners as they turned on their heels. An American flag folded into a triangle, presented to Tracy's widow. She walked past us all, stepping down the aisle with a flag in one hand and an iron rose in the other, the last thing Tracy had ever welded.

Watching the funeral, thinking about Shona waiting for me at home, I couldn't imagine how I would ever slip back into the routines of the firehouse. All of the normal preparations for the day—going over the rig, checking the air bottles—seemed like impotent and absurd gestures. Tracy was a strong fireman, as solid as they come, and yet there was nothing he could have done to save himself. I'm sure he did everything right that day. He checked his equipment in the morning, tightened it all down as the rig swerved through the streets. He must have made a mental size-up of the blaze, picked his spots, and then gone to work fighting a fire he knew he could handle. He was smart and experienced, not likely to do anything reckless or take chances on a lost cause.

Yet nothing had been able to save him. Not equipment. Not experience. Not even the arms of his crew. In the end it was luck—pure, thoughtless, terrible luck—that put Tracy in that spot, probably just inches either way from making it out. More than scared, it made me angry, angry that we could all be so helpless. That a good man like Tracy had given his life for a fire in a worthless building, a building that might still have to be razed despite the effort that the firefighters had put toward saving it.

The last chief processed up the aisle. Tom caught my eye and nodded. I walked over to him, and we shook hands. "Say hi to Shona," he said without emotion. "I haven't seen her in a while." I promised I would and watched him walk off with his head hung, both of us lacking the will for even basic small talk.

I rode the engine back to the firehouse. For a long time, we all sat numb, not talking. When the bell rang, we looked at each other, buttoned up smoke-stained helmets and blackened coats, and drove out the door, sirens crying.

I made a deal when I took this job. I swore an oath to the citizens of Oakland. We're always careful; suicide is not heroic. But the essence of my job will never change. My life for theirs, my health for their safety.

This job of mine is crucial, but also strangely whimsical. No building is worth my life. Yet long after the occupants have evacuated themselves, we often charge into burning buildings in our eagerness to get our shot at the fire. Which is not to say that I quibble with our aggressive style of fire fighting. There's always a kid hiding under a bed or an invalid in a back room until we prove otherwise.

I sometimes wonder if the thrill I get is a selfish thing, if it's fair to the people I love to let myself be swallowed up by uncertainty every time I leave for work. But I can't help myself: I love this. I love the feel of pulling on gloves stiff with ash on the way to yet another fire. I love the sway of the engine, the wind whipping the flag on the back, the sound of the siren. I love the crowds, the people shouting and clapping when the fires go out.

But the truth is that the fires always go out, that with us or without us, they'll never burn forever. The only question is how much of ourselves we will leave there on that day as an exchange, as the price for the fire's going out sooner rather than later. Every fire takes a toll, every fire diminishes the men and women who fight it, men and women who've made the choice. No other job requires such a conscious trade. I've been at this for only five years, and already my neck is so stiff some mornings it hurts to brush my teeth. Sometimes when I go for my daily swim, it feels as if I've been a smoker all my life, and I have to pull up after only a few laps. But I never think about quitting.

And though I barely knew him, I know for certain that Tracy loved it all as well. He must have loved the never-ending nights, the back-breaking labor, the stir-crazy feeling of being locked in one tiny fire-house for days on end. There's nothing quite like the look of envy on the faces of the oncoming shift when you tell them you got your asses kicked last night, three working fires after midnight and the city burning from end to end. I know that Tracy loved every single shift of his career, because I saw a burning building reflected in his eyes once, and I saw him smile when he stepped down off the rig with an ax in his hand.

· 16 ·

See You at the Big One

Insomnia is an odd disease for a firefighter. After a long day of fires and medical runs, firefighters take to their bunks and sleep as diligently as they work. Some of the newer firehouses have private rooms, tiny cubes with paper-thin walls. But many just have bunks, cots laid out in a row with all the luxury of a field hospital. If you're lucky, maybe you get a shoulder-high partition between beds, but in a firehouse you can't expect any more privacy when you're asleep than you get when you're awake. Firemen hit their beds hard but are quick to rise, and we can be into our boots, at the map, on the rig, and out the door within sixty seconds of an alarm bell's coming in.

I'm an insomniac at home, but at work the problem is compounded. I often find myself ghosting around the firehouse late at night, marking the minutes until a run comes in or the new shift shows up to relieve me. I try to sleep, but in the dorm there is too much snuffling and shifting of weight on tired, rusty springs. The guys snore in harmony; when one guy stops to take a breath, somebody will fill the silence, so that the effect is a solid, constant snore, rising and falling without a break.

On shifts when the night noises get too bad, I'll pick up my bedding and head across the apparatus floor to the dayroom, hoping to set up in one of the old chairs. But usually there's someone asleep in there, too, snoring while a late-night infomercial plays on the TV at full volume. In that case I'll open the top compartment of the hose bed on the rig and throw my sleeping bag up there, trying to catch a few minutes of rest by twisting my body into an S curve to snake my spine around the metal couplings.

If I can get over my jealousy at the ability of others to sleep, I actually like the firehouse at 3:00 A.M. I shuffle around from kitchen to watch room to apparatus floor. The engine sits quietly—no lights, no siren—glistening and dark in the half gloom of streetlights through high windows.

The firehouse is never silent, though. A constant low hum comes from the speakers. The refrigerators whine higher and higher, until they shudder into temporary silence and begin their cycle again. Even if they're newly built, all firehouses get old quickly from the constant wear of big men who don't tread lightly. The walls creak and shiver, and the massive belts and gears of the engine take hours to settle to sleep after they've been run. Mice make mad dashes over open ground to pick off the crumbs from dessert. Some nights the hookers outside the windows are loud and boisterous, some nights quietly industrious, and some nights absent altogether. I don't have any idea how they set their schedules. A sleeping firehouse is eerie and just plain wrong, like a roller coaster with no kids on it or a baseball team playing to empty bleachers.

Usually at night I end up in the watch room, where a light is always burning. I check the screen on the dispatch computer for the yellow and blue boxes that indicate who in the city is awake and running. Occasionally at night the screen falls blank. All engines and trucks are accounted for and asleep, and I feel, absurdly, like Oakland's last lookout, the lone protector.

On some nights I sift through the file cabinet looking at old union announcements, emergency logbooks from thirty years ago, and yellowed pictures of long-retired firefighters with giant Afros or immense sideburns. In a closet at Station One, there's a snapshot of Billy that I know is destined to join the pile of those of other old forgotten firefighters. He's in full Class A dress uniform standing in front of a row of fire trucks. He's got a goofy grin on his face, the kind of exaggerated hammy enthusiasm that only he could come up with. He's flashing two thumbs-ups at the camera, and there's a cigarette behind his ear. There's also a black strap across the face of his badge. If you look closely, you can make out the forms of two fire trucks behind him, aerial ladders crossed and hung with an American flag.

The photo was taken the day of Tracy's funeral. Billy must have been about to go in for the service or else just emerged from it. He looks like he always does—clownish, theatrical, unbowed. At first glance the photo seems shocking or disrespectful—how could he be so happy? But if you know Billy, or if you knew Tracy, or if you understand anything about the fire department at all, you'd have to shake your head and smile along with him.

Fighting fire is a profession that can't be modernized. As long as there has been fire, there has been fire out of control. Fire is simple, an acute chemical reaction; it looks careless, but it makes sense. What we do today to stop a fire is the same thing that has always been done, because fire itself will never change. Despite the tools—the Halligan bars and the chain saws, the foam nozzles and the thermal-imaging cameras—we have three basic techniques available to us. Fire can be cooled, smothered, or starved; there are no other options. And of these three, only one is at all practical. Smothering a fire would require that all the oxygen in the air surrounding us be removed, which is clearly impossible. Fortunately, atmospheric oxygen doesn't burn readily, or else we'd be at risk from every spark off a dragging chain and there would be an explosion with every flick of a match. Oxygen nourishes a fire, but it can't act alone. It must have fuel. Wood. Plastic. Clothing. Even steel, if the temperature is high enough. When a house burns, we can't remove the fuel, can't lift the building and move it away from the flames. So all that is left to us is cooling. We spray water on a fire so that the energy of the inferno will be consumed by the work of evaporation. The fire exhausts itself transforming our water into steam. Simultaneously, when firefighters open a hole in a roof, the superheated gases are lifted up, dissipated into the atmosphere, where there is nothing to damage. Somebody once asked me why our helmets look the way they do, with the wide brim extending backward. It's because of the cooling. The water that is not immediately evaporated is boiled instead, and it falls back down on the firefighters. We use the helmet as a shield to protect the thin skin at the back of our necks.

Not much about our job is different from what it was in 1869, when Oakland launched its first professional fire companies. The gear is a little fancier, but there's still no tool that will fight a fire for you. There's no substitute for lying on the floor of a burning building and taking a beating, one inch at a time, until the fire is close enough to hit with your hose stream. In fact, the only significant change has been the efficiency of the water supply, a welcome move from bucket brigades to steel mains and high-pressure pumpers. A fire hydrant is called a "fireplug" from the days when the pipes beneath the streets were made of hollowed logs with wooden stoppers placed at intervals. But an ax is still an ax. We will always wear them at our waists.

"When I was a new kid, all we wore were helmets and cotton sweatshirts," Tom told me one day. "We didn't even use gloves. If it got too smoky, you'd pull your shirt over your face and try to breathe through that, but that was about it. Used to be the only color smoke you'd see was black. Now it's all yellow and green, and who knows what's in it? It's a good thing I'll be gone soon. You new kids are fucked."

I can't imagine fighting a fire without all my protective gear. I lost a glove once, and I felt completely naked, utterly unsafe. The coat, pants, boots, and helmet that the department bought me when I started cost over a thousand dollars for the set. The coats are heavy, made of some space-age fiber. You have to fight your way through a thicket of buckles, snaps, and Velcro to get yourself out of it after a fire. Oakland was one of the last departments in the country to require that firefighters wear the same fire-resistant material on their legs. When I was new, we fought fires in turnout coats and wool pants—the same crisp slacks that we wear with our public dress uniforms. It always seemed odd to me that we took so much care with our upper bodies but left our legs vulnerable in our wools. "Have you ever seen a sheep on fire?" Tom said when I asked him about it.

We had a fire in a cement-and-brick convenience store one night early on in my career, in the "brown-belt" stage, as Tom would call it, when what I thought I knew eclipsed what I actually knew by a wide and dangerous margin. I stretched the hose line to just outside the front door, and as I looked back, Tom was standing in the cone of light from a streetlamp, swinging his arm over his head in a wide circle and calling, "Water! Water! Ready for water!" The truck crew had already forced the door with a few quick swings from a Halligan and an ax, but it had swung back shut. When I pushed the door open, I saw black smoke, thick like mud, rolling out over my head and into the night. Heat waves shimmered in front of my mask. "Don't get in too deep!" Tom shouted.

The fire was in the back, just visible when I lay down and looked along the floor. Boxes were piled along the back wall and were burning freely, slowly spreading to the shelves of Twinkies and Dinty Moore Beef Stew—the almost-food that Oakland liquor stores sell in order to keep their patrons standing long enough to buy another bottle. The fire had probably been smoldering for hours, and the cement had hemmed in a ferocious heat. I started inside with a low duckwalk; Tom crouched just behind me. All at once I felt pain along both thighs, not as if I'd been touched by something hot but as if my legs themselves were searing from within. I dropped to my stomach and started wriggling forward along the floor, dragging the hose behind me until it pulled tight. "Lighten the line!" I called back to Tom, thinking the hose had snagged on a corner. He crawled up to me and tugged at my foot, letting me know it was time to go. "I'm about to get it!" I yelled.

"No you're not!" he shouted back. I tried to fight him for the line, but he shook his head and held fast, pointing toward the door. I took the nozzle and backed out.

"Feel that?" he said, pulling the mask off his face.

"What? I could have gotten it!" I said, pissed that Tom was laughing. I shouldn't have been upset. Tom was always laughing.

He patted the tops of his thighs. "Feel that?" he asked again. "It's too hot. There's nothing in there worth getting burned for."

By this point the truck crew had vented the tar-and-gravel roof, cut a nice four-by-four-foot hole right over the seat of the fire. We masked back up and went in again, almost standing straight up now. With all the heat gone through the roof, we had what was basically a trash fire extinguished and mopped up before the truckies had a chance to come down the ladder and admire their handiwork from the inside.

Tom knew that I was letting my equipment be my safety net, using my gear instead of my brains to evaluate the situation. While the new stuff is good, it creates a danger by separating the firefighter from his fire. Jackets are stronger than the bodies inside them. Firefighters have burned to death inside their coats, too encapsulated to know that they've gone too far. The trend is toward fire-resistant pants and hoods as well as flaps that descend from the helmets to cover the ears. Old firefighters have blistered ears, gnarled scabs just under the hairline. The modern quest for "safety" robs the firefighter of his sense of touch, his early-warning system that he's gone too deep.

· · ·

I once heard a fire chief say in all seriousness that scientists are developing a hockey-puck-size device that will immediately extinguish a burning building. I had an image of firemen trading in the trucks for a fleet of mopeds, on which we could scoot around the city flinging magic disks.

While it seems unlikely that science will ever "solve" fire, it is true that the incidence of structure fires is on the wane. Cigarette smoking is down—good for the world but bad if you like fire. New buildings are put up with sprinklers and fire doors, alarm panels and fireproofing. More and more homes now have smoke detectors and fire extinguishers, which keep small fires small and us at the firehouse. But there will always be fire. As long as there are heat and oxygen and carelessness, as long as there are fuel and flame and poverty, there will always be fire.

In recent years fire departments across the country have been frantically recasting themselves to avoid obsolescence. The buzzwords in fire-management circles are "all-risk," the idea that firefighters should be able to manage all catastrophes, not just fires and medical calls. To that end, most large departments have gotten into the hazardous-materials business, training firefighters to deal with chemical spills and tanker explosions. Other risks like earthquakes, flooding, downed power lines, and terrorism have forced departments to buy specialized gear and send members to odd classes. Since no single firefighter can be expert in everything, we gravitate to our areas of interest and hope to be on duty the day our particular nightmare befalls a citizen of Oakland. I'm a water-rescue guy, still trying to get some use out of my old white-water experience. We don't have much to work with—the lake, the estuary, a few little creeks—but I keep my skills up and wait for my chance.

"All-risk" has also had the unintended consequence of educating the public to view firefighters as handymen with big red toolboxes on wheels. People have come to believe that no emergency is too small to merit a 911 call. I've rolled on busted pipes, clogged toilets, leaking roofs. I've raced a massive fire engine through streets crowded with schoolchildren to find a woman whose emergency was that her television had come unplugged.

When city budgets tighten, the public never wants to see its fire protection cut. Consequently, fire administrators have to play political games to satisfy heads of other departments, as well as city managers who sometimes view firehouses as financial sinkholes. We don't bring in any revenue and are always breaking costly things like ladders, fire engines, and our bodies. So in order to justify our ever-growing budget, we've had to become slicker, more polished, less like cowboys and more like salesmen. The people we help are called "clients" now. Our rescues are referred to as "customer service." It is an odd semantic formulation, since what we do is anything but a business. Our customers have no interest in the product we offer and pray that they'll never give us repeat business. Additionally, we spend a huge portion of our time trying to put ourselves out of our market, donating smoke alarms, checking sprinkler systems, teaching kids how to "stop, drop, and roll."

Additionally, since fire is by definition unpredictable, it defies all attempts at orderly management. Unlike wildfires, urban fires don't have a season. We can't reduce staffing in the winter or offer limited services on the weekends. Around the country this has led to increasing friction between management and frontline personnel. Oakland trucks used to run with six firefighters; now the standard is four, and one of those four is often detailed out for training or public relations. The city saves money on salaries, though firefighter injuries as well as citizen life and property losses increase exponentially with every pair of hands removed from the rig.

In truth, though, the fire department has had no choice but to change; it's impossible to ignore new realities. Fires are down, and every municipal agency has to be accountable for what it does and how. The traditional patterns of hiring and promotion—the sons of white firemen being selected by the sons of white firemen—are unacceptable in a city as diverse and progressive as Oakland. No longer can firemen drink on duty or abuse the power and privileges of the badge. Many firefighters (myself often included) ridicule the new professionalism, disparage the coddling and the handholding we're expected to do. But we're here for the public. Not only do they pay our salaries and our pensions, they also respect and depend on us.

In the new corporatized world of the fire service, we are subject to daily performance evaluations, lectures on team building, and seminars about proper ergonomic typing techniques. I recall sitting in a sexual-harassment class—no small issue in a job that requires men and women to bunk alongside one another—and watching the crusty veterans try to make sense of the new rules of engagement. The instructor, a little nub of a man, fidgeted nervously as he repeated a mnemonic of his own invention that he was obviously quite proud of. It was designed to help us remember what situations would be considered harassment, and it centered on the issue of what a "reasonable person would find offensive." Just as I was beginning to wonder whether any definition of "reasonable" could apply to men who take pride in burn scars, Jack Alvarez spoke up from the back row. He'd spent the meeting silently spitting his tobacco into a Styrofoam cup and staring out the window.

"What about my dick?" he asked. "Is my dick reasonably offensive? I've always thought it looked pretty good."

I can't say for sure if fire fighting attracts a certain kind of person or if it creates one. But I've met firefighters from all over the country, and from other countries, and they're all the same.

Any firefighter would understand that Billy's smile in the photo at Tracy's funeral didn't mean he wasn't grieving. They'd know that he loved Tracy and that losing him opened up a little hole inside him that can't ever be filled, only covered over and stashed away. It's tempting to make someone into a hero after he dies. But Tracy wasn't a hero because he died—he was a hero because he lived. Billy smiles out of the photo because he's proud that Tracy was there to do his job and because there are thousands more like him who are ready to do it again, despite the realities we all know of but rarely speak about. That's why when we put that black strap over our badges, we place it so it covers the name of the city we work for. When a firefighter dies, it doesn't matter what city he worked for, just that he did the work we all do, took the same risks, kissed his sleeping wife good-bye in the morning, and never let himself think for a second that it would be the last time he'd ever do it.

I don't work at Station One anymore. I put in three years there and then moved on. For a job that never really changes, the people who do it are remarkably transitory. People get promoted and go to new firehouses, guys get tired of each other and ready for a change. New firefighters come out of the drill tower all the time, bringing new ideas, new personalities that eventually settle in and find their fit. I've crossed over to the other edge of downtown; my view from the firehouse roof is much the same as it was from Station One, except that I see the skyscrapers from the other side now. Antoine earned a promotion to engineer and then another one to lieutenant. Soon, I'm sure, he'll be a captain and then a battalion chief. Everyone knew he was destined to wear a lot of brass.

I still see the guys from Station One, Tom and Billy and Captain Powell, because our districts overlap and we often meet in the middle. We shake hands at two in the morning after we've silenced a false alarm or gotten a knock on a garbage-chute fire in a tenement. I know that Tom's old-timer friends keep asking him to take it easy, come join them at a slower house up in the hills, but he'll never go. Same with Captain Powell. The two of them are in their element down among the high-rises. They can't help but feel safe in the knowledge that they've worked together for years, that they've each found somebody they can rely on no matter what happens.

Oakland has changed a lot since I came in just a few years ago. The string of lights that runs around Lake Merritt has been relit after having been darkened for years. People jog around the lake at night, not scared of every dark corner anymore. The necklace of lights illuminates their path with greens and blues and reds. Downtown seems more vibrant than when I started. It wasn't long ago that more windows were shuttered than not, and my family would never have considered coming into the heart of Oakland after dark. Now funky restaurants and small boutiques dot every street. The grass in front of city hall is verdant and alive, and businessmen and -women actually come outside for their lunch hour now. They spend money, lounge on the public benches, and enjoy the sunshine. I like parking the fire engine on the street at noontime, or on a warm Friday evening when people are going to the shows and walking along the waterfront. Folks let their kids clamber around in the rig and try on our gigantic coats. We pass out little sticker badges and red plastic hats that say FIRE CHIEF. People aren't so afraid of Oakland anymore.

The word "firefighter" has a strange connotation. It seems like a natural cousin to "crime fighter," but it's different. Fire fighting isn't really about fire in the same way that police work is obsessed with crime. I don't have fire in my blood; I'm not obsessed by it. I respect fire and believe it can be beautiful at times, but I don't lie awake nights thinking of it as my sworn enemy. Fire fighting is not about the thing we conquer but about the things we save. The fire is secondary; the lives are always first. For all our macho posturing, we're just caretakers. Fires, medicals, vehicle accidents—they're really all the same. Firefighters are like mercenaries of rescue; we want to save someone, and we don't much care who the enemy is. Alvarez likes to tell me that I'm different, that I'm too soft to be a real fireman. But I think it's he who can't admit that all the screaming and cursing and breaking things is just a pose to cover up the fact that so much of the job is about taking care of people.

It's hard to be a rescuer, because the opportunity doesn't arise much in normal life. But once you find a niche, and your talent for being there, the act of rescuing becomes instantly addictive. It always feels nice to help someone, but it's also fun to kick the shit out of a door or smash a car windshield with an ax. I don't know many people who quit this job before it's time to retire. The highs are just too high for anyone to want to leave.

But sometimes I hate this job. When my neck hurts, when it feels as if I haven't slept in weeks, when the most exciting thing I've done in a month is roust a sleeping drunk or pour water into a smoldering trash can. On those nights I wonder if I'm wasting my life cooped up inside an ugly little cement firehouse, worrying over strangers, living their lives instead of my own. And firefighters can be an easy target. The public sees us buying ice cream in the store or playing basketball in the backyard. They read about our guaranteed union salaries, and they're jealous of all the time off we get. Twenty-four hours on duty followed by forty-eight hours off—every day we work is Friday. But the repetitiveness of the schedule can be punishing. Like clockwork, my personal life stops for one day out of every three; one-third of my life belongs exclusively to the citizens of Oakland. I'll miss one-third of my anniversaries, one-third of the summer nights I could be spending on the balcony with Shona. If I ever have kids, I'll miss one-third of their soccer games, one-third of their prom nights, one-third of their phone calls home from college.

And it's more than that. Tracy Toomey died one night, within reaching distance of retirement, with a wife, children, and countless firemen who loved him. His partner that day can't walk, sit, or stand without pain. The city retired her and cut their losses. The lieutenant who went inside with them has had countless skin grafts to repair the scalded legs he got from lying under the same burning debris that killed Tracy. He wasn't even forty, and now he'll never fight fire again. Nights like that help you decide whether you fight fire for yourself or not.

I have a photocopy of the investigator's report for the fire that killed Tracy. I've talked over some of the generalities of the fire with people, tried to learn a few concrete lessons that I can take away. But I can't bring myself to pore over the specifics, can't bear to reduce the tragedy to the details of building codes, water-flow rates, and "likely ignition sources." What I know for sure is that saying good-bye to Shona in the morning only gets harder with every shift I work.

I've got a good helmet now. It's dented and filthy, the visor buckled from all the heat it's taken. Some of the neon reflectors have peeled off, and the ones that are left are so soot-blackened as to be useless. I've heard of new kids putting their helmets in the oven or dragging their turnout coats behind their cars so that the gear can get that battle-scarred look. But my stuff has started to look good on its own. I don't know when it happened, but I'm getting to be one of the old-timers now. I'd never say that around Alvarez or Tom, but it's true. A new pension system has made it attractive for a lot of guys to retire years earlier than they ever thought they would. The city has been on a hiring blitz ever since I came in, and there are now 150 guys below me in seniority, over half the force. Most of the nonpromoted guys who have more time than I do have made the decision to head for the hills. In the busy flatlands, firefighters from my generation fill most of the jump seats.

It all seemed so difficult at the time, but I'm only now realizing what a luxury it was to be a new kid. I like the dependence of being told what to do. It was always so simple: Bottle up, step off the rig, look to Captain Powell and say, "What do you need me to do?" It's been easy to follow behind Tom, do what he does, feel confident that he wouldn't put himself someplace stupid. But Tom's getting near the end. He just bought a brand-new Harley, and he's talking about the places he'll ride once he doesn't have to come to work every third day. Captain Powell is cagey, and he won't say what his plans are, but I can't imagine him staying much longer than Tom. I can't imagine either of them without the other.

There have been days recently when the combined experience of everybody on the rig is less than ten years, and we're all looking to one another to have the answers. We had a car fire a while ago, two SUVs that came off the freeway, flipped over, caught fire, and landed on the train tracks. The guy I was working with was fresh from the tower, still looking for his first thrill, and I let him have the nozzle. I knew that, as good as it looked, those two big wrecks shooting flames to the moon, it was still only a car fire. Nothing worth getting possessive about.

Some of the guys I came in with are lieutenants now. They know just as much about fire as I do. Some of them know less. I can't count on them to know reflexively what to do, to point me toward the safe spots. I also know that I'm not a great firefighter myself, not by any means. I don't have a natural gift for being able to predict where a fire is heading or what needs to be done to keep it from going there. Pure instinct is something that can't be developed, no matter how many fires I've been to. Some guys seem to know just what to do the moment they see their first fire. They know which tools to grab, which flames to aim for. They know, without having to be told, where the limits of safety lie. But I'm not like that. I still go blank at every fire when I first step off the rig and size up the scene. I still have that moment in which I've got no idea in the world what it is I'm supposed to be doing.

The difference now is that those moments are getting shorter. And even when I'm in the middle of one, I no longer have the fear that the moment will never end. Now when the lieutenant yells at me to "pull a big line," I'm already doing it. My body's quicker than my mind. My hands know what to do: Get water, go to work, stop the fire.

I wonder if maybe nobody is a natural, whether it's all a matter of how convincingly we each learn to play the role. This isn't like tightening a nut as it floats past on an assembly line. Fire is still the unruly, temperamental boss, and I've never met a firefighter who isn't humble for at least a little while after a tough job. Maybe anyone who says he's a natural is only a liar instead.

I don't kid myself that I rise to the occasion at a fire because I'm particularly brave or selfless or heroic. In the end I go to work because everyone around me is working, and we're all just trying to fill our roles. When I grab the nozzle, it's not out of courage, but because, at that particular time, the nozzle is the tool I've been given as my one small part of a giant team project. I know that if I don't do what I'm supposed to, then nobody else will either. And if I bail out with fear, I will have disappointed these men and women on whom I rely for my safety, these people who, just minutes before the alarm rang, were burning the roast, destroying me at cards, telling me that the new haircut I got looks terrible and that my old piece-of-shit truck looks even worse. I fight fire because afterward I'll be proud of the black streaks on my face and the soot in my ears. I do it because of the calm and the camaraderie of draining hose in the street after the fire is out. I do it because of the laughs and the jokes and the constant handshaking, and because when guys pull away at the end of any job, they always lean out the window of the rig and wave and say, "I'll see you at the big one."

ACKNOWLEDGMENTS

I would like to thank all of the men and women of Local 55 of the International Association of Firefighters for their patience, generosity, and good humor. I would also like to pay my respects to Dave Honegger, Johnny Morgan, Burl Smith, and Mike Triplett, four very good men with whom I had the privilege of working for far too short a time.

This book would not have been possible without the guidance of my agent, Sloan Harris, whose idea this was, and the help of Scott Moyers, my editor at The Penguin Press, who has been unfailingly good-natured and insightful.

There have been many others who have helped at various stages of this process. Kate Galbraith, Michael Goff, Jeffrey Goldberg, Katie Silberman, and Alex Travelli were instrumental in getting the ball rolling. I would like to especially thank Jodi Kantor and her compatriots at *Slate* magazine, where my writing about the fire department first appeared. Kimberly Burns, Katharine Cluverius, Sophie Fels, Pete Lemieux, and Elena Schneider have given me their expertise and hard work, which have been much appreciated.

ACKNOWLEDGMENTS

The book has been vastly improved by the suggestions of many people: Amy Golden, Marcy Gordon, Jeremy Hess, Jodi Kantor, Janet Lafler, Barbara Lewis, Emily Morganti, Lucien Nunn, Dara Raspberry, and Dale Scott. Copy editor Maureen Sugden did amazing work and saved me from much embarrassment.

And finally, I give all of my love to Shona, who makes me happy.